Math
ADVANTAGE

Teacher's Edition

Georgia Test Preparation

CRCT and Stanford 9

Includes Test Taking Tips and
math assessment items for
- **multiple choice format**
- **short answer**
- **extended response**

Grade 5

Harcourt Brace & Company

Orlando • Atlanta • Austin • Boston • San Francisco • Chicago • Dallas • New York • Toronto • London
http://www.hbschool.com

CONTENTS

© Harcourt

Scoring Short Answer Responses

Students can use 3-5 minutes to respond to short-answer test questions. Short-Answer Responses are scored using a rubric. Students can receive partial credit for a partially completed or partially correct answer.

Poor sentence structure, word choice, usage, grammar, and spelling does not affect the scoring of short-answer items, unless communication of ideas is impossible to determine.

Scoring Rubric

Response Level	Criteria
Score 2	**Generally accurate, complete, and clear** ____ All of the parts of the task are successfully completed. ____ There is evidence of clear understanding of key concepts and procedures. ____ Student work shows correct set up and accurate computation.
Score 1	**Partially accurate** ____ Some parts of the task are successfully completed; other parts are attempted and their intents addressed, but they are not completed. ____ Answers for some parts are correct, but partially correct or incorrect for others.
Score 0	**Not accurate, complete, and clear** ____ No part of the task is completed with any success. ____ There is little, if any, evidence that the student understands key concepts and procedures.

Help Students Understand What Scorers Expect

1. Discuss the rubric with students.
2. Have students score their own answer to a practice task, using the rubric.
3. Discuss results. Have students revise their work to improve their scores.

Help students develop proficiency with short-answer questions.

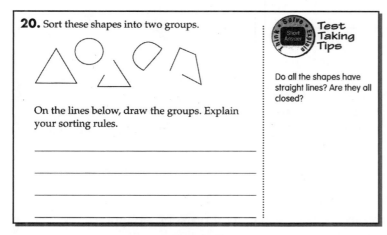

Grade 5

Troubleshooting

Use this discussion to help students answer test items effectively.

"I don't get it!"

Help students read the problem <u>carefully</u>. Then ask, "What do you think you are asked to do?"

"What should I write?"

Have students tell you how they solved the problem. Then have them write their words or use pictures.

"Is this the right answer?"

Have students explain how they know that they have answered the question completely.

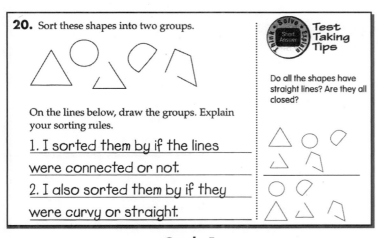

Grade 5

Exemplary response

This student has carefully read the problem. She has shown two sorting rules. She has given a thoughtful explanation.

You may want to make a transparency of this example to share with your students.

Have your students evaluate this response to understand why it is clear and complete.

Scoring Extended Response Items
Performance Task Rubric

Response Level	Criteria
Score 4	**Generally accurate, complete, and clear** ____ All of the parts of the task are successfully completed; the intents of all parts of the task are addressed with appropriate strategies and procedures. ____ There is evidence of clear understanding of key concepts and procedures. ____ Student work and explanations are clear. ____ Additional illustrations or information, if present, enhance communication. ____ Answers for all parts are correct or reasonable.
Score 3	**Generally accurate, with minor flaws** ____ There is evidence that the student has a clear understanding of key concepts and procedures. ____ Student work and explanations are clear. ____ Additional illustrations or information communicate adequately. ____ There are flaws in reasoning and/or in computation, or some parts of the task are not addressed.
Score 2	**Partially accurate, with gaps in understanding and/or execution** ____ Some parts of the task are successfully completed; other parts are attempted and their intents addressed, but they are not successfully completed. ____ There is evidence that the student has partial understanding of key concepts and procedures. ____ Additional illustrations or information, if present, may not enhance communication significantly. ____ Answers for some parts are correct, but partially correct or incorrect for others.
Score 1	**Minimally accurate** ____ A part (or parts) of the task is (are) addressed with minimal success while other parts are omitted or incorrect. ____ There is minimal or limited evidence that the student understands concepts and procedures. ____ Answers to most parts are incorrect.
Score 0	**Not accurate, complete, and clear** ____ No part of the task is completed with any success. ____ Any additional illustrations, if present, do not enhance communications and are irrelevant. ____ Answers to all parts are incorrect.

Help Students Understand What Scorers Expect

1. Discuss the rubric with students.
2. Have students score their own answer to a practice task, using the rubric.
3. Discuss results. Have students revise their work to improve their scores.

Help Students Understand How to Show What They Know

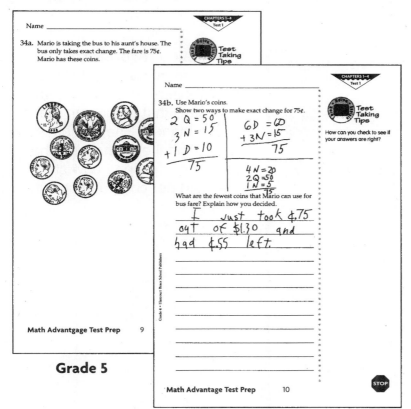

Grade 5

Partially correct response
This student has not read the problem carefully. However, because he did show what he knows about counting money, he will get partial credit for his response.

More practice *reading for information* will help this student read each problem more carefully.

Discussing solution strategies will help him learn to sort through the math knowledge he has to answer the required questions clearly and completely.

Have students evaluate this response to understand how it can be improved.

Ask, "Did the student answer each question?"
Help students see that the student answered the first question but showed more than two ways. The student did not answer the second question. Instead he or she figured out how much change Mario would have left over.

Ask, "What can this student do to improve his or her score?"

The student can restate each part of the problem in his or her own words.
Then the student can check his or her restatement against the problem given and against his or her answer.

Have your students evaluate this response to understand why it is complete.

Name _____

34b. Use Mario's coins.
Show two ways to make exact change for 75¢.

2Q   4N
2D   1Q
1N   3D

Daily Practice
FCAT

Test Taking Tips

How can you check to see if your answers are right?

What are the fewest coins that Mario can use for bus fare? Explain how you decided.

He could use 5 coins with
the ones he has.
There are 25 ¢ in a
Quarter. + 25+25 = 50 There
are 10¢ in a Dime and
10+10 = 20. There is 5¢ in
a Nickle. All together
50 +20 + 5 = 75.

Grade 5

Exemplary response
This student has carefully read the problem. She has shown two ways that Mario can make exact change. She has given a thoughtful explanation. She has also explained how she checked her computation.

Help Students Practice Reviewing and Revising Their Own Work
1. Have a volunteer share a response to a performance task question.
2. Have students discuss the answer.
3. Have students revise their own work to improve their score.

© Harcourt

Name _____

Choose the letter of the correct answer.

1 Yuki has $120.00. He wants to buy a bicycle and gear. The bicycle costs $88.99, the light costs $17.49, and the helmet costs $23.89. About how much more money does Yuki need to buy all the items?

A about $40   C about $20
B about $30   D about $10

Test Taking Tips

Look for important words.
What does "About how much" tell you about the answer?

2 What is the best estimate of the number of eggs?

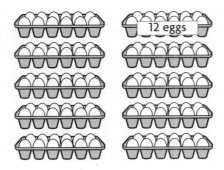

F 12 eggs   H 120 eggs
G 90 eggs   J 1200 eggs

3 What is the value of the digit 4 in the number 207,456,801?

A 4 hundreds
B 4 thousands
C 4 hundred thousands
D 4 millions

4 900
     −428

F 462   H 482
G 472   J 492

5 Jay has 950 marbles. Annette has 576 marbles. How many more marbles does Jay have than Annette?

A 274 more marbles
B 294 more marbles
C 354 more marbles
D 374 more marbles

For questions 6–7, use the table.

Month	Avg. Temp.	Rainfall
May	83°	7.8 in.
June	88°	5.9 in.
July	93°	4.6 in.
August	95°	4.3 in.

6 Which month has the highest average temperature?

F August   H June
G July   J May

7 Of the three months with the lowest temperatures, which month has the lowest rainfall?

A August   C June
B July   D May

GO ON

For questions 8–9, use the table.

POINTS SCORED BY BASKETBALL TEAM			
Game	1st Half	2nd Half	Total
1	39	49	?
2	46	45	91
3	42	51	93
4	?	47	96

8 How many points were scored in Game 1?

F 78 points   H 88 points
G 86 points   J 98 points

9 How many points were scored during the first half of Game 4?

A 45 points   C 51 points
B 49 points   D 55 points

10 What is *thirteen thousand, two hundred ninety* written in expanded notation?

F $13,000 + 200 + 9$
G $13,000 + 200 + 90$
H $1,300 + 200 + 90$
J $1,300 + 20 + 9$

11 $48.307 + 11.6 = n$

A $n = 59.307$
B $n = 59.607$
C $n = 59.907$
D $n = 60.907$

12
$$\begin{array}{r} 15.304 \\ -12.738 \\ \hline \end{array}$$

F 2.466   H 2.566
G 2.476   J 3.566

13 What part of the 10×10 grid is shaded?

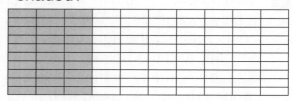

A 0.003   C 0.3
B 0.03   D 3.0

14 Choose the number sentences that can be used to solve the word problem.

A fifth-grade teacher has $85.00 to spend for supplies. He spends $49.95 for construction paper. He still needs to buy glue for $34.29. Does he have enough money?

F $\$85.00 + \$49.95 = n$
G $\$49.95 + \$34.29 = n$
H $\$85.00 - \$34.29 = n$
J $\$85.00 - \$49.95 = n$

15 Estimate the sum to the nearest tenth.

$$\begin{array}{r} 5.383 \\ +3.279 \\ \hline \end{array}$$

A about 8.7   C about 9.0
B about 8.9   D about 9.1

16 Use the number line.

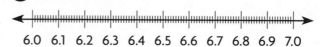

Which number is less than 6.23?

F 6.18   H 6.50
G 6.31   J NOT HERE

© Harcourt

Name _____

17 Samantha will arrive at Mountain Camp on the afternoon of Sunday, August 3. Her parents will pick her up the morning of Saturday, August 16. She has promised to write one postcard each night at camp. How many stamps should she pack?

_____ stamps

Explain your answer on the lines below.

AUGUST						
Su	M	Tu	W	Th	F	Sa
					1	2
3	4	5	6	7	8	9
10	11	12	13	14	15	16
17	18	19	20	21	22	23
24	25	26	27	28	29	30
31						

_____

_____

_____

Test Taking Tips

How can you use the calendar to find the number of stamps?

18 Jennifer, Martin, and Trevor collected aluminum cans for the Recycling Marathon.

They compared the number of cans they collected and wrote this message on the board.

"Trevor collected fewer cans than Jennifer. Jennifer collected more cans than Martin. Martin collected more cans than Trevor."

List the children in order from the one who collected the fewest cans to the one who collected the most cans.

_____

_____

_____

_____

Test Taking Tips

How can drawing a diagram help you solve the problem?

GO ON

19 Nick and his mother bought these things for his trip to camp.

1 jogging suit	$16.99
1 4-pack socks	$9.95
2 T-shirts	$10.95
1 jacket	$13.95

Estimate the total cost of these items.

_____

Explain your method.

_____

_____

_____

Think • Solve • Explain
Short Answer

Test Taking Tips

How will rounding help you solve the problem?

20 Here is a riddle about the ages of people in a family.

Esteban is twice as old as 4-year old Alicia. Juan is 2 years younger than Esteban. The sum of their ages is 8 less than their aunt's age. How old is their aunt? Explain how you found the answer.

_____

_____

_____

_____

_____

Think • Solve • Explain
Short Answer

Test Taking Tips

How is the aunt's age related to the ages of the three children?

GO ON

© Harcourt

Name _____

21 Julia counted the number of nickels in her piggy bank. She can make equal stacks with four nickels in a stack, or five nickels in a stack, or six nickels in a stack. What is the fewest number of nickels that Julia could have counted?

_____

Explain how you solved the problem.

_____

_____

_____

Test Taking Tips

How can you use factors and multiples to help you solve the problem?

22 Rosie is arranging cards to make the least number possible if she does not start with zero.

| 7 | 5 | 0 | 0 | 1 | 9 | 8 |

Write the number: _____

Explain how you find the least number.

_____

_____

_____

Test Taking Tips

What do you know about place value?

© Harcourt

Math Advantage Georgia Test Prep   5   **GO ON**

Name _____

23 What number can I be? I am greater than 3. I am less than 10.

On the number line below, draw dots to show all the whole numbers that I can be.

Test Taking Tips

How can you model the riddle on a number line?

```
←─┼──┼──┼──┼──┼──┼──┼──┼──┼──┼──┼──┼──┼─→
  0  1  2  3  4  5  6  7  8  9  10 11 12
```

24 The fifth grade at Jefferson Elementary School has 300 students. The fourth grade has 242 students. How many more students are in the fifth grade? Check your answer.

Test Taking Tips

How do you subtract across zeros?

_____

_____

_____

_____

_____

_____

_____

_____

_____

_____

GO ON

Name _____

25 **a.** Ms. Young's fifth grade is recycling cans and bottles for cash. They get 3 cents per can and 5 cents per bottle at a local recycling center.

They have collected this many so far:

 Test Taking Tips

Recycling for Cash

Week	Number of Cans	Number of Bottles
Week 1	45	22
Week 2	82	99
Week 3	133	71

How many cans were collected in all?

How many bottles were collected in all?

The class wants to find out which has earned more cash: recycling cans or recycling bottles.

Find the total number collected in all. Show your work in the space below.

number of cans: _____

number of bottles: _____

© Harcourt

25 **b.** Find the money earned by recycling cans and recycling bottles.

Show all your equations.

How can you check that your answers are correct?

Which earned more: recycling cans or recycling bottles?

_____

Explain your work.

_____

_____

_____

_____

_____

_____

_____

© Harcourt

GO ON

Name _____

26 **a.** How many miles do people walk in one year on the job? Here are some statistics.

How Many Miles We Walk

Job Title	Miles Walked in One Year (average)
Police Officer	1,632
Mail Carrier	1,056
TV Reporter	1,008
Nurse	942
Doctor	840

How many miles does a mail carrier walk in 3 years?

_____

_____

_____

_____

_____

_____

_____

_____

_____

© Harcourt

Test Taking Tips

What must you do to determine how many miles they will walk in 3 years?

GO ON

Name _____

26 b. In one year, how many more miles does a police officer walk than a mail carrier?

_____

_____

_____

_____

_____

_____

How many miles does a doctor walk in an average month?

_____

_____

_____

_____

_____

_____

_____

_____

Test Taking Tips

What operation will you use to answer each question?

Math Advantage Georgia Test Prep   10

© Harcourt

Name _____

Choose the letter of the correct answer.

1 516
 $\times$ 8

  A 4,118   C 4,526   E NOT HERE
  B 4,126   D 4,528

Test Taking Tips

Eliminate choices.

If you solve the problem and don't see your solution listed, mark NOT HERE as the answer.

2 $9 \times \underline{\ ?\ } = 9$

  F 0   H 9   K NOT HERE
  G 1   J 81

3 Choose the multiplication property that is used in the number sentence.

  $(4 \times 3) \times 5 = 4 \times (3 \times 5)$

  A Associative Property
  B Property of One
  C Commutative Property
  D Zero Property for Multiplication

4 Find the perimeter of the rectangle. Each square is 1 ft long.

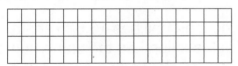

  F 24 ft   H 40 ft   K NOT HERE
  G 32 ft   J 64 ft

5 Susan's bathroom floor is 9 ft long and 8 ft wide. What is the area of the floor?

  A 17   C 64   E NOT HERE
  B 18   D 81

6 54
 $\times$17

  F 808   H 908   K NOT HERE
  G 818   J 918

7 Each bag of apples has 14 apples. How many apples are in 16 bags?

  A 214   C 234   E NOT HERE
  B 224   D 244

8 Estimate the product by rounding each factor to its greatest place-value position.

  297
 $\times$ 11

  F 2,600   H 3,000   K NOT HERE
  G 2,800   J 3,800

9 136
 $\times$403

  A 54,798   C 54,890   E NOT HERE
  B 54,808   D 54,898

10 What is the area of the figure?

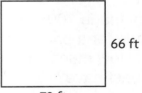

66 ft

78 ft

  F 142 sq ft   H 5148 sq ft   K NOT HERE
  G 144 sq ft   J 5248 sq ft

11 6)426

  A 61   C 70 r2   E NOT HERE
  B 64   D 72

Math Advantage Georgia Test Prep   11   GO ON

© Harcourt

12 Find the volume of the box.

Use the formula $V = l \times w \times h$.

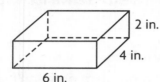

2 in.
4 in.
6 in.

F 24 cu in. **H** 96 cu in.
G 48 cu in. **J** 128 cu in.

13 8)617

A 76 **C** 77
B 76 r2 **D** 77 r1

14 8)550

F 68 r6 **H** 69 r6
G 68 r7 **J** NOT HERE

15 There are 846 students in a middle school. Each of the 3 grades has an equal number of students. How many students are in each grade?

A 258 students
B 282 students
C 292 students
D NOT HERE

16 Solve the problem and decide how to interpret the remainder.

Mrs. Fry made 100 glasses of lemonade for a party. There are 30 guests. How much lemonade can each guest have?

F 3 glasses; drop the remainder.

G $3\frac{1}{3}$ glasses; use the remainder as part of the answer.

H 4 glasses; round the quotient to the next greater number.

17 The fifth-grade classes found 246 shells. They divided them equally among all students and had 6 leftover. How many students took shells home? How many did they take?

A 25 students took 9 shells each
B 28 students took 7 shells each
C 29 students took 6 shells each
D 30 students took 8 shells each

18 $180 \div 30 = n$

F $n = 0.60$ **H** $n = 60$
G $n = 6$ **J** $n = 600$

19 Estimate the quotient.
52)4,978

A about 80 **C** about 800
B about 100 **D** about 1,000

20 18)435

F 24 r1 **H** 24 r3
G 24 r2 **J** 24 r6

21 A concert hall has 16,625 seats. Each section has the same number of seats. How many seats are in each of the 35 sections?

A 425 **B** 450 **C** 475 **D** 490

22 Choose the number sentence used to solve the problem.

Mrs. Roberts saved $5,880 last year. She saved the same amount each month for 12 months. How much did she save each month?

F $\$5,880 \times 12 = n$
G $\$5,880 \div 12 = n$
H $\$5,880 + 12 = n$
J $\$5,880 - 12 = n$

GO ON

Name _____

23 Assume you eat 3 meals a day, 7 days week. Estimate how many meals you will eat in one year.

_____

On the lines below, explain how to estimate the number of meals you will eat in 10 years.

_____

_____

_____

_____

_____

_____

_____

Test Taking Tips

Short Answer
Think • Solve • Explain

What number will you use to estimate the number of weeks in a year?

24 Anna's allowance is $5 a week. She can earn more by doing extra chores. She can earn $1 per day for walking the dog and $1 per day for doing the dishes.

How much can she make in four weeks if she does every chore?

_____

Show how you figured it out.

Test Taking Tips

Short Answer
Think • Solve • Explain

How much can she earn in one week?

© Harcourt

GO ON

25 The Youth Group is organizing for a trip for 250 people, including the adult supervisors. They will be going in vans. Each van will be driven by one adult and will carry 6 passengers.

Will 30 vans be enough for 250 people? Explain your answer.

_____

_____

_____

_____

_____

_____

Test Taking Tips

What operation can you use to solve the problem?

26 Mr. Ellis has 36 students in his classroom. He wants to arrange their desks in rows, with an equal number of students in each row. How many ways can Mr. Ellis arrange the desks in his room?

_____ rows, _____ seats in each row

_____ rows, _____ seats in each row

_____ rows, _____ seats in each row

Test Taking Tips

What are the factors of 36?

© Harcourt

GO ON

27 Frank's car is a gas guzzler.

He recently went 180 miles on 15 gallons of gas.

Helen owns a fuel-efficient car.

She recently went 180 miles on 9 gallons of gas.

How far does Frank's car go on 1 gallon of gas? _____

How far does Helen's car go on 1 gallon of gas? _____

On the lines below, compare the gas mileage of the two cars. Explain the measurement you use to compare them.

_____

_____

_____

Test Taking Tips

What operations can you use to solve the problem?

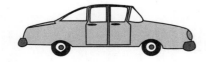

---

28 Shana has a new bead kit. It has 240 beads in it. She wants to use all the beads. She wants to make all the necklaces the same. Which design can she choose and have no beads left over?

    Design A: uses 24 beads

    Design B: uses 36 beads

    Design C: uses 16 beads

Explain how you decided.

_____

_____

_____

Test Taking Tips

What operation can you use to solve the problem?

Name _____

29 George is flying 2,690 miles from San Jose, California, to Boston, Massachusetts. The flight takes about 5 hours.

Boston ←

San Jose →

What is the airplane's average rate of speed per hour?

_____

Explain how you solved this problem.

_____

_____

_____

_____

Test Taking Tips

What operation can you use to solve this problem?

30 Police Officer Early walks, on the average, 1,650 miles a year. He works 50 weeks a year. How many miles a week does he walk?

Write your answer in the space below. Show all your work.

_____

_____

_____

_____

_____

_____

Test Taking Tips

What operation can you use?

GO ON →

31 **a.** Here is a riddle about the ages of people in a family.

Stan is twice as old as 5-year-old Alex. Ben is 2 years younger than Stan. The sum of their ages is 10 less than their dad's age.

How old is their dad?

_____

_____

_____

_____

_____

_____

_____

_____

_____

_____

_____

_____

_____

_____

Test Taking Tips

How can you compare Stan's age to Alex's age?

How can writing number sentences help you solve the problem?

GO ON

Name _____

31 **b.** Explain your solution strategy.

_____

_____

_____

_____

_____

_____

_____

_____

_____

_____

_____

_____

_____

_____

_____

_____

_____

_____

How can you check that your answer is accurate?

How can you check that your explanation is clear and complete?

GO ON ➡

Name _____

32 **a.** This is a diagram of Sea Breeze Stadium. It has seats for 48,000 people. Each section has the same number of seats.

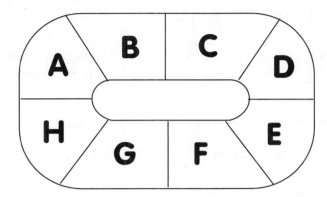

About how many seats are in each section?

_____

Explain how you found out.

_____

_____

_____

_____

_____

_____

_____

_____

Test Taking Tips

What operation can you use to solve the problem?

GO ON

Name _____

32 **b.** The diagram shows how full each section was at a recent concert.

Section A	Section B	Section C	Section D	Section E	Section F	Section G	Section H
almost full	almost full	full	full	half full	mostly empty	mostly empty	half full

About how many people attended the concert?
Estimate to the nearest 1,000.

_____

_____

Explain your method.

_____

_____

_____

_____

_____

_____

_____

_____

_____

_____

Think • Solve • Explain
Long Answer

Test Taking Tips

How many seats are in each section?

STOP

Name _____

Choose the letter of the correct answer.

1 What is the graph designed to show?

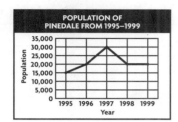

A number of people in Pinedale
B number of babies born in Pinedale
C population changes over 5 years
D NOT HERE

Test Taking Tips

Solve • Explain • Think
Multiple Choice

Decide on a plan.
What can you learn from the title of the graph and the labels on the axes?

For questions 2–3, use the graph.

MONEY SPENT FOR TENNIS EQUIPMENT	
Item Bought	**Decimal Part**
Racket	0.5
Tennis Balls	0.2
Tennis Shirt	0.2
Headband	0.1

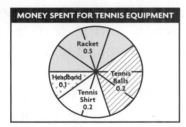

2 Which fraction shows the amount spent for the tennis shirt?

F $\frac{1}{10}$ **G** $\frac{2}{10}$ **H** $\frac{4}{10}$ **J** $\frac{5}{10}$

3 Why isn't the graph correct?

A The labels *Headband* and *Tennis Shirt* are reversed.
B The *Tennis Shirt* part has the wrong number of sections.
C The labels *Tennis Shirt* and *Racket* are reversed.
D One section for *Tennis Balls* should be with *Racket*.

4 Steve can buy black or brown shoes. The shoes are sandals, hiking boots, or tennis shoes. From how many possible shoe combinations can Steve choose?

F 3 possible combinations
G 5 possible combinations
H 6 possible combinations
J 8 possible combinations

5 On the line graph below, the symbol ⤶ is used. What does this symbol mean?

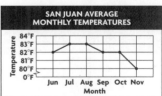

A The range is great.
B The interval is large.
C The data have been averaged.
D There is a break in the scale.

6 What graph would be best for comparing favorite pizza toppings?

F bar graph   **H** line graph
G circle graph   **J** NOT HERE

GO ON

For questions 7–8, use the table.

FOOD SURVEY	
Favorite Types of Food	Number of Students
Pizza	50
Hamburgers	20
Fries	30
Hot Dogs	10

7 What is the most reasonable scale to use to graph the set of data?

A 0, 1, 2, 3, 4, 5
B 0, 3, 6, 9, 12, 15, 18
C 0, 5, 10, 15, 20, 25
D 0, 10, 20, 30, 40, 50

8 How many more students like pizza than like hot dogs?

F 10 more students
G 20 more students
H 30 more students
J 40 more students

9 A bag has 12 blue and 3 red cubes inside it. Picking a red cube is __?__.

A likely   C certain
B not likely   D impossible

For questions 10–11, use the circle graph.

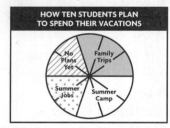

10 Of the 10 students, what decimal part plan to have summer jobs?

F 0.1   G 0.2   H 0.3   J 0.4

11 Which fraction represents the students going on family trips?

A $\frac{5}{10}$ B $\frac{3}{10}$ C $\frac{2}{10}$ D $\frac{1}{10}$

For questions 12–13, use the stem-and-leaf plot.

High Temperatures in May (°F)

Stem	Leaves
6	0 1 2 3 3 5 6 6 7 8 9
7	0 1 1 2 3 4 4 5 5 7 8 9
8	0 3 4 4 5 5 6 8

12 To find the mean high temperature for May, you add the temperatures and divide by __?__.

F the middle number
G the highest number
H the number of days
J NOT HERE

13 What is the median for the set of data?

A 73°   B 75°   C 84°   D 88°

14 Megan has 6 triangles, 2 circles, 3 rectangles, and 18 squares in a box. Which shape is she most likely to pick?

F square   H rectangle
G circle   J triangle

15 In September, 915 students attended Evergreen Middle School. In October, 904 students attended. In November, 918 attended. In December, 890 attended, and 922 attended in January. What graph would be best to display this data?

A circle graph
B bar graph
C line graph
D stem-and-leaf plot

GO ON

Name _____

16 Mr. Cohen's class made this graph in early December. Each student was asked, "During the last six months, when did you buy shoes?" If a student bought shoes two different months, both months were put in the data.

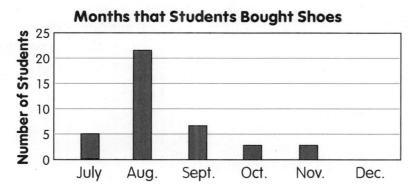

Months that Students Bought Shoes

In which month did the students buy the most shoes?

_____

On the line below, write about two things you can say based on the graph.

_____

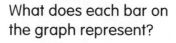
Test Taking Tips

What does each bar on the graph represent?

17 Janet said, "The graph shows that most people surveyed do NOT prefer pizza." Is Janet correct?

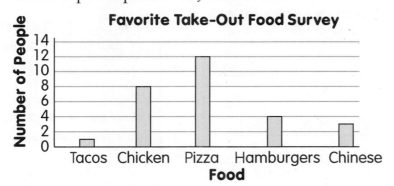

Favorite Take-Out Food Survey

Explain how you decided.

_____

_____

Test Taking Tips

How many people are represented in the survey?

© Harcourt

Math Advantage Georgia Test Prep   23

GO ON

Name _____

18 About half of babies born are boys and about half are girls. Here is a tree diagram showing the possible combinations of boys and girls in a family with 2 children.

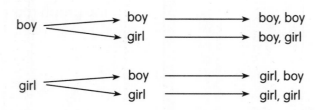

First Child   Second Child   Children in Family

Given this information, is it more likely that a family will have 2 girls, 2 boys, or 1 girl and 1 boy?

_____

On the lines below, explain how you decided.

_____

_____

Test Taking Tips

Does any combination of possibilities appear more than once?

19 All the fifth grade students at Carmen's school voted on the theme for Spirit Day. Each student had one vote.

Votes on Theme for Spirit Day	
Silly hats	𝍷𝍷𝍷
Presidents	𝍷𝍷
Sports	𝍷𝍷𝍷
Inventions	𝍷𝍷

How many students voted?

_____

Test Taking Tips

How many votes were cast for each choice?

GO ON

Name _____

20 Cheryl and her younger sister are playing a spinner game. Cheryl gets 2 points if the pointer stops on an even number. Her sister gets 2 points if the pointer stops on a number greater than 4.

Complete each statement.

It is certain that _____

because _____

It is impossible that _____

because _____

It is likely that _____

because _____

Test Taking Tips

How many numbers on the spinner are odd numbers?

How many number on the spinner are greater than 4?

21 Lillian has trouble deciding what to wear. She made two spinners to help her decide. The PANTS spinner has three equal sections: red, white, and blue. The SHIRTS spinner has four equal sections: red, white, blue, multicolor. How many outfits can she make?

Explain how you decided.

_____

_____

_____

_____

_____

Test Taking Tips

How can making a tree diagram help you solve the problem?

GO ON

Name _____

22 There are five errors in Henry's graph of his marble collection. List the errors on the lines below.

Henry's Marble Collection	
Blue	5
Red	3
Green	1
Swirl	4

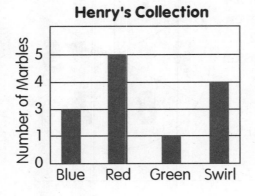

Henry's Collection

_____

_____

_____

Test Taking Tips

How can you compare the data in the table to the data in the graph?

23 Maria went shopping for skates. Here are the prices she found.

$99 $120 $155 $60 $119

What is the median price of the skates? _____

On the lines below, explain how you found the median.

_____

_____

_____

_____

Test Taking Tips

What is the median of a set of data?

GO ON

24 a. The three Morrison children are named Annabelle, Bart, and Casey. All three children have visited Cousin Zack who came down with a cold.

It is possible that Annabelle, Bart, and Casey will get sick. It is possible that none of them will get sick. Many other combinations are possible, such as: one could be sick and two could be well.

Annabelle made a tree diagram to try to understand it. (She abbreviated names with A, B, and C.)

Complete the tree diagram.

How many possible outcomes are there?

Test Taking Tips

How can you use a chart to help summarize the possible outcomes?

Possible Outcomes

well	sick
A, B, C	---

```
                              C is well
              B is well  <
                              C is sick
A is well  <
                              C is well
              B is sick  <
                              C is sick

                              C is well
              B is well  <
                              C is sick
A is sick  <
                              C is well
              B is sick  <
                              C is sick
```

_____

_____

© Harcourt

24 b. How many outcomes are possible if Annabelle stays well? List the outcomes.

_____

_____

_____

_____

_____

_____

_____

_____

_____

_____

_____

_____

_____

_____

_____

_____

Test Taking Tips

How can you check that your results are correct?

GO ON

Name _____

 25 a. Colin is arranging his collection of sports cards into small boxes. Count the boxes and fill out the following data chart.

Think • Solve • Explain
Long Answer
Test Taking Tips

How can you change fractions to percents?

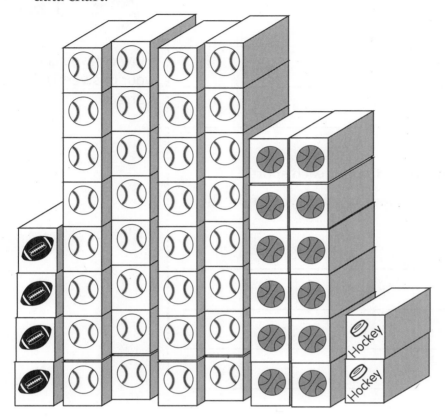

Complete the table. Show the number of boxes for each kind of sports card in Colin's collection.

Colin's Sports Card Collection

	Football	Baseball	Basketball	Hockey	TOTAL
Number of Boxes					

Name _____

25 b. What fraction of the collection is each kind of card?

football _____

baseball _____

basketball _____

hockey _____

Find the percent of the collection for each sport.

Explain how you can find percent when you know the fraction.

_____

_____

_____

_____

_____

_____

_____

_____

_____

_____

_____

Test Taking Tips

How can you check that your answers are correct?

Name _____

Choose the letter of the correct answer.

1 Use mental math to complete the pattern.

$11 \times 6 = 66$

$11 \times 0.6 = 6.6$

$11 \times 0.06 = n$

A $n = 0.066$ **C** $n = 6.6$

B $n = 0.66$ **D** $n = 66$

Test Taking Tips

Understand the problem.

Look for patterns in the second factor of the multiplication equations.

2 Choose the best strategy to solve this problem.

$1 \times 35 = 35$

$0.1 \times 35 = n$

$0.01 \times 35 = 0.35$

F paper and pencil   **H** mental math

G calculator   **J** a ruler

3 $0.5 \times 0.9 = n$

A $n = 0.45$ **C** $n = 4.0$

B $n = 1.4$ **D** $n = 4.5$

4 Which would you use to estimate $68 \times \$3.14$?

F $60 \times \$3$ **H** $70 \times \$4$

G $65 \times \$3$ **J** $70 \times \$3$

5 $0.003 \times 17 = n$

A $n = 0.00051$ **C** $n = 0.051$

B $n = 0.0051$ **D** $n = 0.51$

6 Potatoes are on sale for $0.69 a pound. What is the price of 6.8 pounds of potatoes to the nearest dollar?

F $6.00   **H** $4.00

G $5.00   **J** $3.00

7 Which equation has a decimal in the quotient?

A $9 \div 5 = n$

B $90 \div 5 = n$

C $900 \div 5 = n$

D $9{,}000 \div 5 = n$

8 Complete the pattern.

$900 \div 30 = 30$

$90 \div 30 = 3$

$9 \div 30 = n$

F $n = 0.003$ **H** $n = 0.3$

G $n = 0.03$ **J** $n = 30$

9 The Halls have subscribed to an Internet service. It costs $540.00 for 12 months. How much does it cost each month?

A $35.00   **C** $45.00

B $40.00   **D** $50.00

10 $4\overline{)0.52}$

F 0.13   **H** 13.0

G 1.3   **J** 130

© Harcourt

GO ON

11 If a walnut weighs 9 grams, how much does it weigh in milligrams?

A 90 mg   C 9,000 mg
B 900 mg   D 90,000 mg

12 What is the most reasonable unit of measure for finding the mass of a horse?

F milligram   H kilogram
G gram   J meter

13 $27.6 \div 3 = \underline{}$

A 0.92   B 8.9   C 9.2   D 92

14 A bag of 8 cookies costs $1.52. How much does each cookie cost?

F $0.20   H $0.18
G $0.19   J $0.17

15 Sticks of cotton candy at the zoo cost $0.95 each. Blake bought 7 sticks. How much did they cost him?

A $7.35   C $6.55
B $6.65   D $6.35

16 The zoo gives its elephant 16 pounds of food per day. How many pounds of food does the elephant get in one week?

F 72 pounds   H 106 pounds
G 96 pounds   J 112 pounds

17 You estimate how long a rug is by walking off its length. What is a reasonable estimate?

A 8 millimeters   C 8 inches
B 8 centimeters   D 8 feet

18 Choose the missing unit of measurement to complete the equation.

87 dm = 870 $\underline{}$

F mm   H dm
G cm   J km

19 Before going on a hike, 5 people divided 3 liters of water equally. How much water did each person get?

A 60 mL   C 660 mL
B 600 mL   D 6,000 mL

20 Clara drinks 275 mL for each mile she bikes. If she bikes 7 miles, how much water does she drink?

F 1,925 L   H 19.25 L
G 192.5 L   J 1.925 L

21 Complete the multiplication sentence for the drawing.

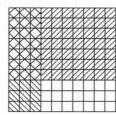

$n \times 0.7 = 0.21$

A $n = 0.03$ C $n = 3$
B $n = 0.3$ D $n = 30$

22 Choose the missing unit of measurement to complete the equation.

560 cm = 56 $\underline{}$

F mm   H dm
G cm   J km

© Harcourt

GO ON

Name _____

23 Charmaine is a bargain hunter. She finds SuperFine hair scrunchies for $0.89 a piece.

Make a table. Show how much Charmaine would pay for 8, 12, 16, and 20 SuperFine hair scrunchies.

Charmaine's SuperFine Hair Scrunchies

Number	4				
Cost	$3.56				

Test Taking Tips

What operation can you use to solve the problem?

24 Sheila likes to shop at the mall. She wants to estimate the distance she walks from one store to another. Name at least two units of measurement that Sheila is likely to use.

_____

_____

Explain your choices.

_____

_____

_____

_____

Test Taking Tips

How will she measure a distance of five or ten steps?

Name _____

25 This is a life-size picture of a ruby-throated hummingbird.

If you want to find a more precise measurement from the tip of its beak to the tip of its tail, which units of measure should be used?

Choose millimeters or centimeters, or inches or feet.

Explain your reasoning.

_____

_____

_____

_____

Test Taking Tips
Think • Solve • Explain
Short Answer

Will you use smaller units or larger units of measure?

26 Angelo and Martina are solving number pattern riddles. Find the next two numbers in this pattern.

2, 6, 18, 54, _____ , _____

Describe the pattern that helped you decide.

_____

_____

_____

_____

_____

Test Taking Tips
Think • Solve • Explain
Short Answer

What operation can you use to find the next number using the number before?

GO ON

Name _____

Name _____

27 Alana wants to arrange the plants in her garden in two or more rows with an equal number of plants in each row. Can she do it if she has 19 plants?

Write your answer on the lines below. Explain how you decided.

_____

_____

_____

_____

_____

Is 19 a prime or a composite number?

28 Match each phrase to the correct numerical representation.

1. a little less than half   a. 0.52

2. almost none   b. $\frac{9}{10}$

3. almost one   c. 0.03

4. a little more than half   d. three out of eight

What benchmark decimals and benchmark fractions can help you solve the problem? How can reading each number aloud help you decide?

Math Advantage Georgia Test Prep   35

© Harcourt

Name _____

29 All cars need engines to make them go. Some countries such as the United States measure engine capacity in horsepower. Other countries measure engine capacity in cubic centimeters (cc).

Study the table below to show how the two measures of engine capacity are related. Use the Jaguar E-car data to find the relationship.

Engine Capacity

Name of Car	Engine Capacity (in cubic centimeters)	Engine Capacity (in horsepower)
lightest car	2.5	
Mazda RX2		90
AMC Hornet		128
Volvo Sedan	1,780	
Ford Thunderbird		300
Jaguar E-car	4,235	423.5

Write the rule for the relationship between cubic centimeters and horsepower.

_____

 Test Taking Tips

Compare the two measures of engine capacity for the Jaguar. What ratio do you see?

30 Sharon and Mandy were buying candy for a party. Sharon selected chocolate covered raisins that cost $2.95 per pound. Mandy selected jelly beans that cost $0.25 per ounce. Each girl bought two pounds of candy. Who spent more?

On the lines below, explain your thinking.

_____

_____

_____

Test Taking Tips

What information do you need to find out how much money Mandy spent?

 © Harcourt

31 **a.** Sara is going to replace the tile on her bathroom floor. She wants to estimate the cost of doing the whole floor. She is going to start by figuring how much it would cost to tile one sample square 15 inches on each side.

SQUARE FLOOR TILES

Tile Flooring	Length of each Side	Price
White	5 inches	$1.76 each
	3 inches	$0.67 each

The squares below represent Sara's sample square that is 15 inches on each side.

Sketch a plan with 3-inch square tiles in this sample.

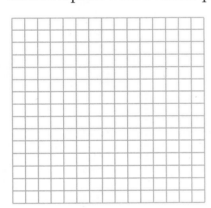

_____ 3-in. tiles

Sketch a plan with 5-inch square tiles in this sample.

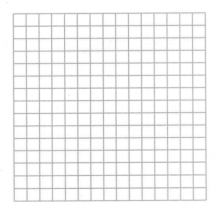

_____ 5-in. tiles

How does drawing a model help you solve the problem?

GO ON

31 b. How much would it cost to tile the sample square with 3-inch tiles?

_____

How much would it cost to tile the sample square with 5-inch tiles?

_____

Which tiles should Sara use to save money?

Explain how you found out. Show your equations.

_____

_____

_____

_____

_____

_____

_____

_____

_____

_____

_____

Test Taking Tips

How can drawing a diagram help solve the problem?

© Harcourt

GO ON

Name _____

32 a. Brenda is buying food for a picnic. She can buy hotdogs in packages of 10 for $1.99, hotdog rolls in packages of 8 for $0.99, and juice in cases of 24 cans for $3.99.

She has $50.00 to spend. She has invited 48 people. Help Brenda make a shopping list.

She wants to serve each person two hotdogs, each on a roll, and two cans of juice. How much food should she buy? Use the table below to make Brenda's shopping list.

Brenda's Shopping List

_____

_____

_____

_____

_____

_____

_____

Test Taking Tips

Think • Solve • Explain

Long Answer

What will each person be served at the picnic?

GO ON →

Name _____

32 **b.** Use estimation. Decide whether Brenda will be able to
stay within her budget.
You may add to the table in Part A.
Explain how you solved the problem.

_____

_____

_____

_____

_____

_____

_____

_____

_____

_____

_____

_____

_____

_____

Test Taking Tips

How can rounding help
you solve the problem?

How can you check that
your explanation is clear
and complete?

STOP

Name _____

Choose the letter of the correct answer.

1 What is $\frac{8}{3}$ written as a mixed number?

A $2\frac{2}{3}$

C $3\frac{2}{3}$

B $3\frac{1}{3}$

D NOT HERE

Test Taking Tips

Decide on a plan.

What two steps are needed to write a fraction as a mixed number?

2 Compare the fractions, using the LCM.

$$\frac{3}{4} \bullet \frac{4}{5}$$

F $\frac{3}{8} < \frac{4}{8}$

H $\frac{15}{20} < \frac{16}{20}$

G $\frac{9}{12} > \frac{8}{12}$

J NOT HERE

3 $\frac{7}{8} \bullet \frac{8}{9}$

A $<$ B $>$ C $=$

4 A spinner has 10 equal sections. The spinner is $\frac{5}{10}$ red, $\frac{2}{10}$ yellow, $\frac{1}{10}$ blue, and $\frac{2}{10}$ green. How much of the spinner is covered with yellow and green?

F $\frac{2}{10}$

H $\frac{2}{5}$

G $\frac{3}{10}$

J $\frac{4}{5}$

5 What is the greatest common factor for the pair of numbers?

28 and 70

A 18   B 17   C 14   D 13

6 Use the number lines to name an equivalent fraction for $\frac{4}{5}$.

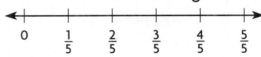

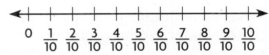

F $\frac{2}{10}$ G $\frac{4}{10}$ H $\frac{5}{10}$ J $\frac{8}{10}$

7 Gabrielle has 12 cousins. Of her cousins, 4 are girls. What fraction of Gabrielle's cousins are girls?

A $\frac{1}{3}$ B $\frac{3}{9}$ C $\frac{9}{12}$ D $\frac{3}{4}$

8 Of 60 people at a theater, 25 went to see a science fiction movie. In simplest form, what fraction of the people at the theater went to see the science fiction movie?

F $\frac{4}{5}$ G $\frac{3}{4}$ H $\frac{2}{3}$ J $\frac{5}{12}$

9 Use an inch ruler to find the difference.

$$\frac{5}{16} \text{ in.} - \frac{1}{8} \text{ in.} = n$$

A $n = \frac{1}{16}$ in.   C $n = \frac{1}{4}$ in.

B $n = \frac{3}{16}$ in.   D $n = \frac{1}{3}$ in.

© Harcourt

Math Advantage Georgia Test Prep   41   **GO ON**

10 Robin ate $\frac{1}{8}$ of a round of cheese. Her sister ate another $\frac{1}{8}$ of the cheese. How much of the cheese did they eat altogether?

F $\frac{5}{6}$ of the cheese

G $\frac{1}{2}$ of the cheese

H $\frac{1}{3}$ of the cheese

J $\frac{1}{4}$ of the cheese

11 What is the least common denominator for the group of fractions?

$$\frac{2}{5} \quad \frac{1}{2} \quad \frac{3}{10}$$

A tenths   **C** thirds
B sevenths   **D** NOT HERE

12 Bill walks $\frac{1}{2}$ mile every day. How far does he walk in 5 days?

F $6\frac{1}{2}$ miles   **H** $3\frac{1}{4}$ miles

G $5\frac{1}{2}$ miles   **J** $2\frac{1}{2}$ miles

13 Use fraction strips to find the sum expressed in simplest form.

$$\frac{3}{8} + \frac{3}{8} = \underline{\quad ?\quad}$$

A $\frac{3}{8}$ **C** $\frac{5}{8}$

B $\frac{1}{2}$ **D** $\frac{3}{4}$

14 Alicia found $\frac{9}{12}$ of a chocolate pie left in the pan. She ate $\frac{1}{3}$ of the remaining pie. How much of the pie was left after she ate a piece?

F $\frac{1}{8}$ **G** $\frac{1}{4}$ **H** $\frac{1}{3}$ **J** $\frac{1}{2}$

15 Use fraction strips to find the difference expressed in simplest form.

$$\frac{4}{5} - \frac{1}{2} = \underline{\quad ?\quad}$$

A $\frac{1}{10}$ **B** $\frac{1}{5}$ **C** $\frac{3}{10}$ **D** $1\frac{1}{5}$

16 Mary has $\frac{3}{16}$ pound of flour and $\frac{1}{2}$ pound of sugar. How much more sugar than flour does she have?

F $\frac{1}{16}$ pound   **H** $\frac{5}{16}$ pound

G $\frac{1}{4}$ pound   **J** $\frac{11}{16}$ pound

17 Choose the fraction shown on the fraction strip.

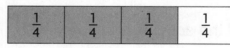

A $\frac{1}{4}$ **B** $\frac{1}{3}$ **C** $\frac{2}{3}$ **D** $\frac{3}{4}$

18 On Tuesday a plant measured $\frac{7}{8}$ in. On Monday it had grown $\frac{1}{2}$ in. How tall had the plant been before it grew on Monday?

F $\frac{3}{8}$ in.   **G** $\frac{1}{2}$ in.   **H** $\frac{5}{8}$ in.   **J** $\frac{7}{8}$ in.

© Harcourt

19 Here is a picture of the prize-winning pie before the judges tasted it.

Judge Hogg ate one-third of the pie. Judge Lite ate one-sixth of the pie. Judge Mello ate one-fourth of the pie.

How much of the pie is left?

On the lines below, explain how you decided.

_____

_____

_____

_____

_____

Test Taking Tips

How can you use equivalent fractions to solve the problem?

---

20 Dana's mother baked an apple pie. Dana ate $\frac{1}{3}$ of the pie. Dana's father ate $\frac{2}{6}$ of the pie. Who ate more pie? How do you know?

_____

_____

_____

_____

_____

_____

Test Taking Tips

How can you compare fractions?

GO ON ➡

21 About one out of 10 people is left-handed.

Mr. Kant, the art teacher, is buying scissors for a school of 100 students. About how many left-handed scissors should he buy? Explain your answer.

_____

_____

_____

Think • Solve • Explain Short Answer

Test Taking Tips

What fraction of all people do left-handed people represent?

22 Tracy has 20 balloons that are red, green, or yellow. If $\frac{2}{5}$ of the balloons are red and $\frac{1}{2}$ of the balloons are yellow, how many balloons are green?

_____

_____

_____

_____

_____

_____

_____

_____

_____

Think • Solve • Explain Short Answer

Test Taking Tips

How do you find a fraction of a number?

GO ON

Name _____

23 To make his favorite pizza, Darrell needs $2\frac{1}{2}$ cups of mozzarella cheese, $1\frac{3}{4}$ cups pizza sauce, $1\frac{2}{3}$ cups mushrooms, and $\frac{1}{3}$ cup pepperoni. Which ingredient has the greatest amount? Which ingredient has the least amount?

_____

_____

_____

_____

Test Taking Tips

How can you order mixed numbers?

24 For the party, Clayton spends $\frac{5}{16}$ of his money on decorations and $\frac{3}{8}$ of his money on food. On which item does he spend more money? Explain how you know.

_____

_____

_____

_____

_____

_____

Test Taking Tips

How can you compare fractions?

© Harcourt

Math Advantage Georgia Test Prep   45   **GO ON**

Name _____

25 Every day that Kevin walks home from school, he walks $1\frac{1}{2}$ miles. He walked home 3 days this week. How far did he walk? Make a model to help you solve the problem.

_____

_____

_____

_____

_____

Solve
Think · Explain
Short Answer

Test Taking Tips

How will making a model help you add fractions?

26 Jennica's pencil was $5\frac{3}{4}$ inches long. After two weeks, it was $5\frac{1}{3}$ inches long. How much of her pencil was gone after two weeks?

_____

_____

_____

_____

_____

Solve
Think · Explain
Short Answer

Test Taking Tips

How do you subtract fractions that have unlike denominators?

Math Advantage Georgia Test Prep 46

GO ON

Name _____

27 **a.** Lisa kept track of her activities in one eight-hour school day.

She made this table.

Lisa's School Day

Activity	Time Spent	Fraction of School Day	Fraction in Simplest Form
Recess	1 hour	$\frac{1}{8}$	$\frac{1}{8}$
Class Learning Time	5 hours	$\frac{5}{8}$	
Homework	2 hours		

Complete the table to show the fraction of a day that Lisa spent in each activity.

Test
Taking
Tips

Think • Solve • Explain
Long Answer

What is the number of hours in Lisa's school day?

GO ON

Name _____

27 **b.** Make a circle graph showing how she spent her day. Be sure to label all the parts of your graph and write a title for it.

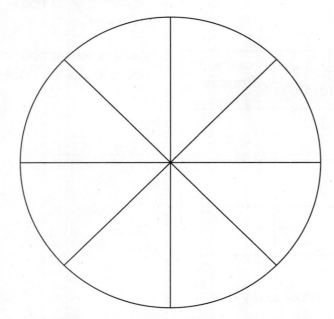

On the lines below, write two statements about how Lisa spent her day. Use data from the graph to support your statements.

1) _____

_____

_____

2) _____

_____

_____

Test Taking Tips

What can you see clearly from the circle graph?

Think • Solve • Explain
Long Answer

GO ON

28 **a.** There are 30 students in Mr. Nguyen's fifth grade. Each student brought one book to donate to the school library.

Complete the table.

Library Book Donations

Type of Book	Number of Books Donated	Fraction of Books Donated
Sports	9	
Humor	3	
Mystery	6	
Adventure	6	
History	3	
Science	3	

Test Taking Tips

How many books were donated altogether?

GO ON

Name _____

28 b. Represent the data on a circle graph.

Be sure to

- write a title for your graph
- label all the parts

Test Taking Tips

Think • Solve • Explain

Long Answer

What fraction of the circle graph does each book type represent?

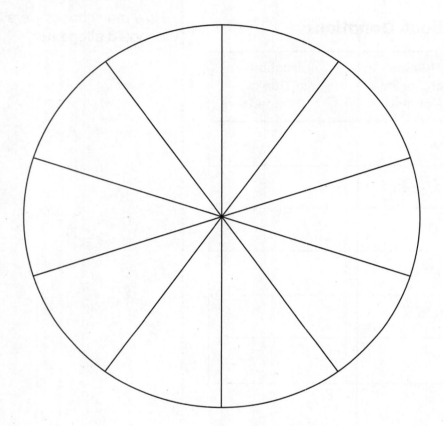

Use your graph to write two statements that compare library book donations.

1)

_____

_____

2)

_____

_____

STOP

Name _____

Choose the letter of the correct answer.

For questions 1–2, find the sum or difference in simplest form.

1 $\frac{7}{8} + \frac{5}{8} = n$

A $n = \frac{1}{4}$　　　**D** $n = 1\frac{1}{2}$

B $n = 1$　　　**E** NOT HERE

C $n = \frac{12}{8}$

2 $\frac{5}{6} - \frac{2}{6} = n$

F $n = \frac{1}{2}$　　　**J** $n = 1\frac{1}{6}$

G $n = \frac{2}{3}$　　　**K** NOT HERE

H $n = \frac{7}{6}$

3 Danielle practiced piano for $\frac{3}{4}$ hour in the evening and $\frac{1}{2}$ hour in the morning. How long did she practice?

A $1\frac{1}{2}$ hour　　　**C** $\frac{2}{3}$ hour

B $1\frac{1}{4}$ hour　　　**D** $\frac{1}{4}$ hour

4 Find the sum.

$$3\frac{1}{5}$$
$$+2\frac{1}{10}$$

F $5\frac{7}{10}$　　**H** $5\frac{3}{10}$　　**K** NOT HERE

G $5\frac{1}{2}$　　**J** $5\frac{1}{4}$

 Test Taking Tips

Decide on a plan.
How do you add fractions with like denominators?

5 Estimate the sum.

$$9\frac{1}{8} + 2\frac{1}{10}$$

A about 7　　　**D** about 19

B about 11　　　**E** NOT HERE

C about 12

Choose the number sentence that goes best with the picture.

6

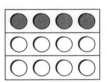

F $8 + 4 = 4$　　　**J** $\frac{1}{4} \times 12 = 3$

G $\frac{1}{2} \times 12 = 6$　　　**K** NOT HERE

H $\frac{1}{3} \times 12 = 4$

7 The temperature at 10:00 A.M. was 65°F. The temperature was 15 degrees lower at 8:00 P.M. By 11:00 P.M. the temperature had dropped 13 more degrees. What was the temperature at 11:00 P.M.?

A 35°F　　**C** 47°F　　**E** NOT HERE

B 37°F　　**D** 67°F

© Harcourt

GO ON

8 $\frac{2}{3} \times 15 = n$

F $n = \frac{2}{3}$ **H** $n = 10$

G $n = \frac{3}{4}$ **J** $n = 12$

9 Kim spent $3\frac{1}{2}$ hours at the mall. She spent $\frac{1}{4}$ of the time eating lunch. How long did she spend eating lunch?

A $\frac{2}{3}$ hour   **C** $\frac{7}{8}$ hour

B $\frac{5}{6}$ hour   **D** $1\frac{1}{4}$ hour

10 Use the number line to estimate whether $\frac{2}{3}$ is closer to 0, $\frac{1}{2}$, or 1.

F 0   **G** $\frac{1}{2}$ **H** 1

11 Subtract. Choose the answer in simplest form.

$$4\frac{5}{6} - 2\frac{1}{3}$$

A $\frac{2}{3}$ **C** $2\frac{5}{6}$

B $2\frac{1}{3}$ **D** NOT HERE

12 Find the difference in temperature.

low $-10°C$, high $15°C$

F 5°   **G** 10°   **H** 15°   **J** 25°

13 Find the product in simplest form.

$\frac{2}{3} \times \frac{5}{6} = n$

A $n = \frac{1}{6}$ **C** $n = \frac{5}{9}$

B $n = \frac{1}{3}$ **D** $n = \frac{4}{5}$

14 Change the unit. You may use a calculator.

4 yd = __?__ ft

F 3   **G** 6   **H** 9   **J** 12

For questions 15–16, use the calendar.

May

Su	M	T	W	Th	F	Sa
					1	2
3	4	5	6	7	8	9
10	11	12	13	14	15	16
17	18	19	20	21	22	23
24	25	26	27	28	29	30
31						

15 Rachel and her mom left on May 8 at 9:00 P.M. for vacation. They reached San Francisco on May 11 at 10:00 A.M. How many days and hours did it take them to get to San Francisco?

A 2 days 11 hours
B 2 days 12 hours
C 2 days 13 hours
D 3 days 13 hours

16 Rachel and her mom plan to stay in San Francisco for 2 weeks. When will they leave?

F May 25   **H** May 18
G May 23   **J** May 4

© Harcourt

GO ON

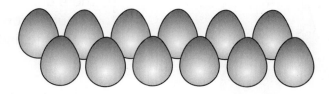
17 Sara is going to color 12 eggs. She wants $\frac{1}{3}$ of the eggs to be blue, $\frac{1}{4}$ to be yellow, and $\frac{5}{12}$ to be green.

How many eggs will she have for each color?

Explain how you decided.

_____

_____

_____

_____

Test Taking Tips

How can you use equivalent fractions to solve the problem?

18 The Ramirez family is having a big reunion. Esperanza estimates that each person will drink about one 8-ounce cup of lemonade. How many people can be served using 10 gallons of lemonade?

On the lines below, explain how you solved the problem.

_____

_____

_____

_____

_____

Test Taking Tips

How many cups are in one gallon?

GO ON

Name _____

19 A backpacker has decided to carry no more than 1 pound of items for recreation on his hike. His son gives him a yo-yo that weighs 3 ounces.

Item	Weight
harmonica	4 ounces
book	6 ounces
camera	13 ounces

Which additional items can be taken without going over the weight limit?

_____

Explain your answer. Use numbers and words.

_____

_____

Test Taking Tips

There are 16 ounces in 1 pound. How will you use this to find an answer?

20 Juanita spent a total of 870 minutes working on her report about the United States Congress.

How many hours did she spend?

_____

Explain how you decided. Use words and equations.

_____

_____

_____

_____

_____

Test Taking Tips

There are 60 minutes in 1 hour.

What operation can you use to convert from minutes to hours?

GO ON

21 Icebergs are formed when parts of a glacier break off where the glacier meets the open sea. Icebergs have about $\frac{3}{4}$ of their height hidden below the surface of the ocean.

This iceberg towers 150 meters above the ocean surface.

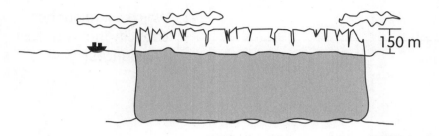

150 m

What is the total height of the iceberg?

On the lines below, explain how you solved the problem.

_____

_____

_____

Test Taking Tips

How can drawing a model help you solve the problem?

---

22 Jerome and Clara want to make a bird house. They need 7 ft of wood. Jerome has $5\frac{1}{2}$ ft of wood, and Clara has $1\frac{1}{3}$ ft of wood. Do they have enough wood? Explain your answer.

_____

_____

_____

_____

_____

_____

Test Taking Tips

How do you add mixed numbers?

GO ON →

23 Sharifa is baking cookies. The recipe calls for 2 cups of flour. She can't find a measuring cup. She has a gallon jug, a juice glass, a teaspoon, and a mixing bowl. Which of these items would be the best substitute for a measuring cup?

_____

Remember: 1 cup = 8 ounces

Explain your answer.

_____

_____

_____

_____

_____

Test Taking Tips

Which container has a capacity that is closest to 8 ounces?

---

24 Chris needs to be $4\frac{1}{2}$ ft tall to ride the roller coaster. Right now he is 50 in. tall. How many more inches does Chris need to grow to ride the roller coaster? Explain how you found your answer.

_____

_____

_____

_____

Test Taking Tips

How can you change inches to feet?

GO ON

Name _____

25 **a.** Jerome is the king in a school pageant.

He is going to wear a cape. When laid out flat on a table, the cape looks like this.

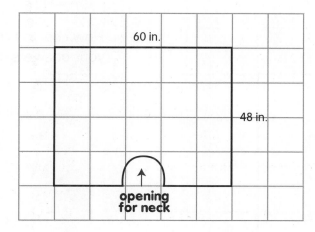

60 in.

48 in.

opening
for neck

How many yards of 60-inch-wide fabric are needed to make the cape? (1 yard = 36 inches). Show your work in the space below.

_____

_____

_____

_____

_____

The teacher has a jar of gold paint that can be used to decorate the cape. The paint will cover an area 12 inches by 12 inches.

How much area does the gold paint cover?

_____

Test Taking Tips

Long Answer

How can you use the diagram to help solve the problem?

GO ON

Name _____

25 b. Add gold to the cape. Color your gold design on to the drawing of the cape. Use some or all of the gold paint in your design.

60 in.

48 in.

↑
**opening
for neck**

Explain your work in the space below.

_____

_____

_____

_____

_____

_____

_____

_____

Test
Taking
Tips

Think • Solve • Explain
Long
Answer

How can you check that
your answers are
correct?

© Harcourt

GO ON ➡

26 **a.** Scott knows that 16 ounces of water weigh 1 pound. He can win a 1-lb bowl of toys if he can balance it with a bowl of water. He can use water from the faucet, a 7-ounce glass, and a 5-ounce glass.

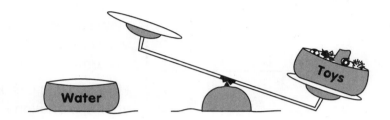

Test Taking Tips

Think • Solve • Explain
Long Answer

What is the difference in the capacity of the two glasses?

Explain how Scott can balance the scale by putting exactly 16 oz of water in the water bowl.

Scott can measure with these water glasses. He can use as much water as he likes.

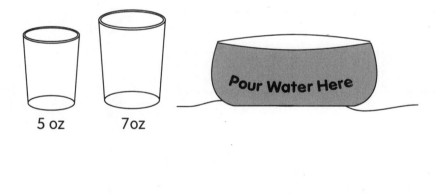

5 oz   7 oz   Pour Water Here

_____

_____

_____

_____

_____

GO ON

26 b. Explain how Scott can use the tools available to fill the water bowl with exactly 16 ounces of water.

_____

_____

_____

_____

_____

_____

_____

_____

_____

_____

_____

_____

_____

_____

Test Taking Tips

How can you check that your answer is accurate?

How can you check that your explanation is clear and complete?

STOP

Choose the letter of the correct answer.

For questions 1–2, use the figure.

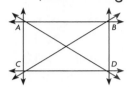

1 Which line segment is parallel to CD?

A AB B AD C BC D AC

Test Taking Tips

Look for important words.

Think about what the word "parallel" means. Draw a picture if that would help you.

---

2 Which line segment is perpendicular to AB?

F AC G AD H CD J BC

For questions 3–4, use the figure.

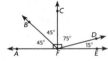

3 Which is an acute angle?

A ∠AFD B ∠CFE
C ∠BFE D ∠BFC

4 Which is an obtuse angle?

F ∠AFB H ∠DFE
G ∠CFE J ∠AFD

5 Use the benchmark to estimate the volume.

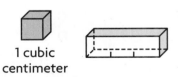

1 cubic centimeter

A about 1 cu cm
B about 3 cu cm
C about 6 cu cm
D about 12 cu cm

6 Which quadrilateral has 2 pairs of congruent sides and 2 pairs of parallel sides?

F parallelogram H triangle
G pentagon   J trapezoid

7 How many lines of symmetry does the figure have?

6 ft   6 ft
6 ft

A 0 lines of symmetry
B 1 line of symmetry
C 2 lines of symmetry
D 3 lines of symmetry

8 Use the figure on the coordinate grid.

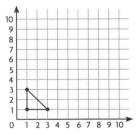

Translate the figure 3 spaces to the right and 4 spaces up. What is the ordered pair of the highest point of the triangle in the new location?

F (3,6) H (4,7) G (6,8) J (6,3)

9 Which of the figures will not tessellate?

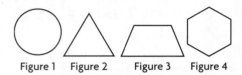

Figure 1   Figure 2   Figure 3   Figure 4

A Figure 1   **C** Figure 3
B Figure 2   **D** Figure 4

10 Which two figures are congruent?

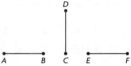

F $\overline{AB}$ is congruent to $\overline{CD}$.
G $\overline{AB}$ is congruent to $\overline{EF}$.
H $\overline{CD}$ is congruent to $\overline{EF}$.

11 Which describes the faces of a triangular pyramid?

A 4 triangles, 1 square
B 5 triangles
C 2 triangles, 2 rectangles
D 4 triangles

12 Identify the solid figure.

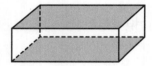

F rectangular prism
G hexagonal pyramid
H triangular pyramid
J rectangular pyramid

13 A circle is divided into 4 sections. Three of the angles measure 90°, 135°, and 45°. What does the fourth angle measure?

A 270° **B** 180° **C** 125° **D** 90°

14 Which of the boxes can hold the most?

F 12 in. × 6 in. × 6 in.
G 10 in. × 8 in. × 10 in.
H 10 in. × 6 in. × 12 in.
J 12 in. × 10 in. × 8 in.

15 Which of the expressions explains how the value of π is found?

A $r + d$ **C** $C \div d$
B $d - r$ **D** $C \div r$

16 What is the line segment in circle C called?

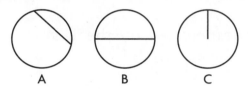

A   B   C

F radius
G diameter
H circumference
J altitude

17 What is the missing angle?

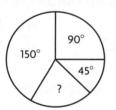

A 45°   **C** 85°
B 75°   **D** 125°

18 Mrs. Wilson has a flower pot with a diameter of 12 inches. What is the radius of the flower pot?

F 1 in.   **H** 6 in.
G 3 in.   **J** 12 in.

GO ON

Name _____

19 Ashley built a house with polygons.

Make a list of the polygons that Ashley used.

_____

_____

_____

Describe the congruent polygons you find.

_____

_____

_____

_____

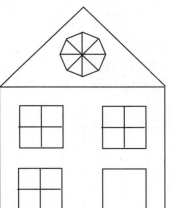

Test Taking Tips

What makes each type of polygon unique?

20 Using the graph below, plot and label points A, B, C, and D. Connect the dots in alphabetical order using 4 line segments. Your last line will return to the first dot.

Point	Coordinates
A	(1,2)
B	(3,4)
C	(5,2)
D	(4,1)

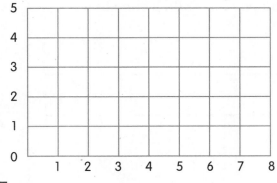

What kind of polygon did you draw?

_____

Test Taking Tips

How do you identify points for ordered pairs in a grid?

GO ON

Name _____

 21 Here are some clues about a geometric figure.
It is a closed figure.
It is two-dimensional.
This figure has 4 straight sides.
Pairs of opposite sides are parallel.

Draw the figure.

On the line below, write a mathematical name for the
shape you have drawn.

_____

Test Taking Tips

What math words will
help you understand the
clues?

22 These blocks are the first three cubes in a pattern of
cubes.

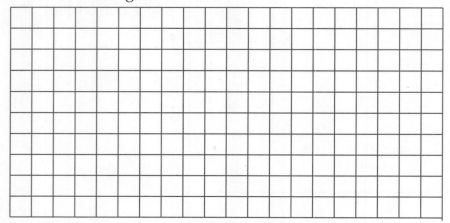

How many blocks will you need to build a cube that
measures 10 blocks on a side?

_____

_____

_____

_____

Test Taking Tips

What do you know
about computing
volume?

Name _____

23 Logan Elementary is building a new playground that can be used by students in wheelchairs. What is the area of the new playground?

Explain how you decided.

_____

_____

_____

Remember: Area is the inside region of a figure.

Find each measurement and show your work.

New Playground

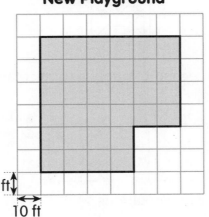

10 ft

10 ft

_____

_____

24 Plot these 4 points on the grid:

(3, 1)

(9, 1)

(9, 5)

(3, 5)

How many square units are enclosed by these coordinates?

_____

_____

_____

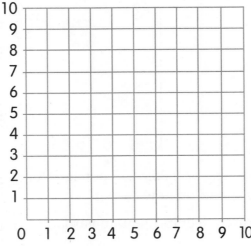

25 John rotated this figure 90 degrees clockwise.

Draw the figure after a second rotation, a third rotation, and a fourth rotation.

Describe the figure after four rotations.

_____

_____

26 Mrs. Nguyen's living room is 12 feet wide and 15 feet long.

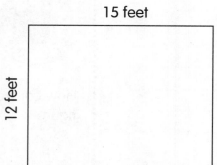

15 feet

12 feet

Her new wall-to-wall carpeting will cost $15 per square yard. How much will she spend on new carpet?

Remember: There are 9 square feet in 1 square yard.

_____

_____

GO ON

27 **a.** The fifth grade has planted gardens at the school. The gardens include three small flower gardens and one large vegetable garden. A map of the gardens is shown below.

4 ft

4 ft

4 ft

4 ft

4 ft

4 ft

Flower Gardens

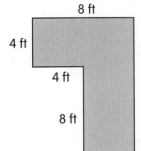

8 ft

4 ft

4 ft

8 ft

**Vegetable
Garden**

Test
Taking
Tips

Think • Solve • Explain
Long
Answer

How many 4-foot
sections are there?

The students have discovered that they need to build a fence around each garden to keep out the woodland animals. Fencing is sold in 8-foot sections. The sections can be cut.

How many 8-foot sections do they need to buy for each flower garden? Show how you decided.

_____

_____

_____

How many 8-foot sections do they need to buy for the vegetable garden? Show how you decided.

_____

_____

GO ON

Name _____

27 **b.** Show how much fencing you need to buy for all the gardens.

Order Form	Number of 8-Foot Sections
Flower Garden	
Vegetable Garden	
Total	

On the lines below, explain how you know that your order is accurate.

_____

_____

_____

_____

_____

_____

_____

_____

_____

_____

Test Taking Tips

How can you check that your answers are correct?

GO ON

28 **a.** Here is a 1 x 1 square. ■

What is its perimeter? _____

Draw the following squares on the grid.

1 by 1, 2 by 2, 3 by 3, 4 by 4, 5 by 5, 6 by 6

Test
Taking
Tips

How are the sizes of the squares related?

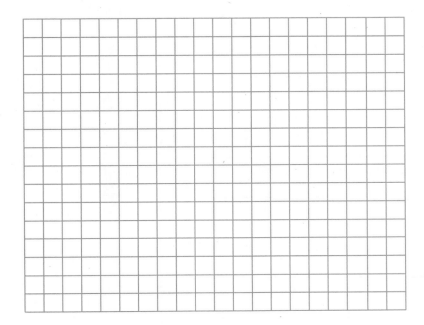

Find the perimeter and area of each square. Record your findings in the table.

Kind of Square	Perimeter	Area
1 x 1 square		
2 x 2 square		
3 x 3 square		
4 x 4 square		
5 x 5 square		
6 x 6 square		

GO ON

Name _____

28 **b.** On the lines below, describe patterns you see in the table. What is the relationship between the area of a square and its perimeter?

_____

_____

_____

_____

_____

_____

_____

_____

_____

_____

_____

_____

_____

_____

_____

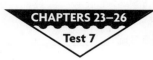

Test Taking Tips

How can you check that your answers are correct?

STOP

Name _____

Choose the letter of the correct answer.

For questions 1–2, find the length of the missing side in the similar shapes.

1

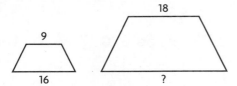

   A 16   **B** 24   **C** 28   **D** 32

2

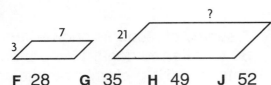

   F 28   **G** 35   **H** 49   **J** 52

For questions 3–4, use the table.

PITCHES THROWN IN ONE INNING	
Number of Pitches	Type
12	Fastball
7	Curve
4	Slider
5	Change-up

3 What is the ratio of curves to fastballs?

   A $\frac{4}{12}$ **B** $\frac{7}{12}$ **C** $\frac{12}{7}$ **D** $\frac{12}{4}$

4 What is the ratio of sliders to change-ups?

   F 4:5   **H** 5:4
   G 7:4   **J** 4:7

5 There are 7 soccer players on the team who have new uniforms. The team has 18 players. What is the ratio of players with new uniforms to the whole team?

   A 18:11   **C** 11:7
   B 18:7   **D** 7:18

Think • Solve • Explain • Multiple Choice

Test Taking Tips

Get the information you need.

Consider how the lengths of the sides of the two figures are related.

For questions 6–7, choose the equivalent ratio.

6 15:45
   F 1:5   **G** 5:5   **H** 5:9   **J** 5:15

7 $\frac{2}{5}$

   A $\frac{4}{10}$ **C** $\frac{8}{16}$

   B $\frac{49}{100}$ **D** $\frac{5}{6}$

For questions 8–9, use the drawing of a farm and the scale.

Crops on a Farm

Farm House		Wheat
Yard	Driveway	
Potatoes		Corn

Scale:
1 linear unit = 12 ft
1 square unit = 144 sq ft

8 What is the length of the driveway?
   F 12 ft **G** 48 ft **H** 84 ft **J** 96 ft

9 What is the area of the cornfield?
   A 432 sq ft   **C** 4,032 sq ft
   B 4,002 sq ft **D** 4,402 sq ft

GO ON ➡

10 What percent of the large rectangle is shaded?

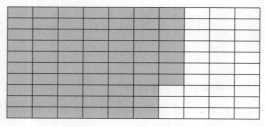

F 0.67%   H 67%

G 6.7%   J 670%

11 Ron put together a model car. In the picture on the box, the car was 2 in. wide and 5 in. long. When the model was assembled, it was 4 in. wide. How long was it?

A 5 in.   C 12 in.

B 8 in.   D NOT HERE

12 What is 0.72 written as a percent?

F 0.72%   H 72%

G 7.2%   J 702%

13 What is 25% written as a fraction in simplest form?

A $\frac{2.5}{100}$

B $\frac{25}{100}$

C $\frac{1}{4}$

D $\frac{1}{2}$

14 Kim's scout troop wanted to sell 100 boxes of cookies on Saturday. They sold 56 boxes during the morning. What percent did they have left to sell?

F 40%   H 50%

G 44%   J NOT HERE

For questions 15–16, choose the benchmark percent to estimate each percent.

15 46%

A 25% B 50% C 75% D 100%

16 68%

F 25% G 50% H 75% J 100%

17 A survey of students showed that 80% preferred dogs, 15% preferred cats, and 5% preferred rabbits as pets. What decimal represents the students who prefer cats?

A 0.05   C 0.55

B 0.15   D 0.80

18 Use the table.

FAVORITE ACTIVITY	
Activity	**Percent of Votes**
Playing Sports	30%
Watching TV	15%
Reading	15%
Visiting with Friends	25%
Doing Hobbies	15%

Which three choices received 45% of the vote?

F watching TV, reading, and visiting with friends

G watching TV, reading, and doing hobbies

H reading, playing sports, and doing hobbies

J playing sports, visiting with friends, reading, and watching TV

GO ON

Name _____

19 On a map, the scale shows that 1 inch = 100 miles. Jeremy is planning a trip that measures $3\frac{1}{4}$ inches on the map. Jeremy thinks the distance is 400 miles. Is he correct? Explain your reasoning.

_____

_____

_____

_____

_____

_____

Test Taking Tips

How can you use the scale to help you solve the problem?

20 The Cougars football team has a 6 to 2 win-loss ratio. The Mustangs football team has a 12 to 4 win-loss ratio. Are the two ratios equivalent? Explain your answer.

_____

_____

_____

_____

_____

Test Taking Tips

How can you determine if ratios are equivalent?

GO ON ➡

21 Sarah collected flower and animal stickers. In the first week, she collected 4 flower stickers and 2 animal stickers. This is a 2 to 1 ratio. After 7 weeks Sarah had collected 108 flower stickers and 36 animal stickers. Is this still a 2 to 1 ratio? Explain.

_____

_____

_____

_____

_____

Test Taking Tips

How can you determine if two numbers have a 2 to 1 ratio?

22 Find the length of the missing side in the similar figures.

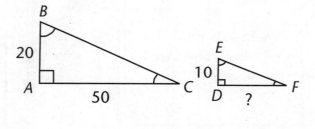

_____

_____

_____

_____

_____

Test Taking Tips

How can you use ratios to find a missing length in similar figures?

Name _____

23 In a clothing store, Mary found a $25 blouse marked "10 percent off." How can she find the reduced cost of the blouse?

_____

_____

_____

_____

_____

_____

Test Taking Tips

How do you find a percent of a number?

24 Mr. Wolfe took a survey of his class. In his class, 25% of the students have 1 pet, 15% have 2 pets, and 10% have more than 2 pets. What percent of his students do not have any pets? Explain how you found the solution.

_____

_____

_____

_____

_____

_____

Test Taking Tips

What should be the sum of all the percents? How will this information help you solve the problem?

GO ON

25 At the water park, the ratio of lifeguards to swimmers is 1 to 15. There are 25 lifeguards on duty. How many swimmers are there? Explain how you found the solution.

_____

_____

_____

_____

_____

_____

Solve • Explain • Think

Short Answer

Test Taking Tips

How can you use a ratio table to help you solve the problem?

26 Ben wants to know how far volleyball practice is from his house. The scale on Ben's map indicates that 1 in. = 3 miles. Ben measures the distance on the map to be $4\frac{1}{2}$ in. How far does Ben travel to go to volleyball practice?

_____

_____

_____

_____

_____

_____

Solve • Explain • Think

Short Answer

Test Taking Tips

How can you use a scale on a map to help you find the actual distance?

GO ON

Name _____

 27 **a.** Mr. Rocket's class conducted a survey of 100 fifth graders. Here are the survey results.

Hours of Television Watched in a Week

Number of Hours	Number of Students
0	5
1–6	35
7–12	40
13 or more	20

What fraction of the students surveyed watches between 1 and 12 hours of television? Explain how you found this answer.

_____

_____

_____

Write a percent for each of the answers given in the survey.

0 hours _____   1–6 hours _____

7–12 hours _____   13+ hours _____

Which answer was the most popular? least popular?

_____

_____

_____

Test
Taking
Tips

How can you compare the fractions and percents?

GO ON

Name _____

 b. Use the information gathered by the students to make a circle graph.

Be sure to

• write a title for your graph.

• label all the parts.

How do you use percents to help create a circle graph?

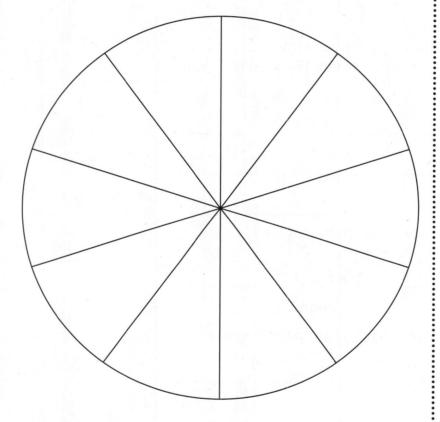

GO ON

Name _____

28 a. Dan wants to make a scale drawing of his room at home. Dan decides that the scale for his drawing is going to be 2 ft = 1 in.

He measures his room at home. The length of the room is 12 ft and the width is 16 ft. He also measures some other things in his room. He organized his measurements in the chart below.

Object	Measurement
Bed	4 ft X 6 ft
Dresser	4 ft X 2 ft
Desk	6 ft X 4 ft

Find the length and width of his room using the scale 2 ft = 1 in.

_____

_____

Using the scale find the dimensions for each of the objects that Dan measured in his room. Place your answers in the table.

Object	Measurement	Scale Measurement
Bed	4 ft X 6 ft	
Dresser	4 ft X 2 ft	
Desk	6 ft X 4 ft	

Test Taking Tips

How can you use a ratio to help you solve the problem?

© Harcourt

GO ON

28 b. Dan's room is a rectangle. He plans to make his scale drawing on a piece of paper that is $8\frac{1}{2}$ in. $\times$ 11 in. Is the drawing going to fit on the paper? Explain your answer.

_____

_____

_____

_____

_____

_____

_____

_____

_____

_____

_____

_____

Think • Solve • Explain
Long Answer

Test Taking Tips

How can you determine if the scale drawing will fit on the paper? What information do you need to use?

STOP

Georgia Quality Core Curriculum Objectives

5.1 Rounds whole numbers to nearest ten, hundred, or thousand; decimals to the nearest tenth or whole number; and fractions to the nearest whole number.

5.3 Use estimation strategies (such as front-end, clustering, rounding, or reference point) to predict computational results of whole numbers, fractions, mixed numbers, and decimals.

5.5 Relates decimals (through hundredths) to models using base ten blocks and grid paper.

5.6 Using models and vertical and horizontal presentations with the horizontal rewritten vertically, adds and subtracts decimals through hundredths, without and with regrouping.

5.16 Determines and estimates amounts of money.

5.17 Identifies different names for numbers (whole numbers, fractions, and decimals) including number words and expanded notation and relates models to such numbers.

5.18 Identifies place value for whole numbers through millions, and decimals through hundredths. Determines the effect that changing a digit will have on the value of the number.

5.22 Compares and orders whole numbers, fractions, and decimals through hundredths.

5.24 Identifies needed information and selects the steps necessary to solve multi-step word problems.

5.25 Solves one-, two-, and three-step word problems related to all appropriate fifth grade objectives including those presented orally and in writing; those in charts, tables, and graphs; and those with extraneous or insufficient information.

5.30 Interprets and draws conclusions from charts, tables, and graphs (e.g., pictographs, and circle graphs).

5.36 Adds, subtracts, multiplies and divides whole numbers up through four digits using both vertical and horizontal (rewrite vertically) presentation without and with regrouping (uses calculators for more laborious computations than these: four-digit addition and subtraction; multiplication and division of three- by two-digit numbers).

1. Answer: D about $10

Discussion

Tip: What does the phrase "About how much" tell you about the answer?

Students should recognize that these words signal an estimate. Help students sequence the steps needed for the estimate. Ask:

Which numbers need to be rounded?
What is the approximate cost of the 3 items?
How much greater is this than $120?

You may want to discuss the mental math strategies needed for each step.

Item Numbers	Georgia QCC Objectives
1. D	5.1, 5.16
2. H	5.3
3. C	5.18
4. G	5.36
5. D	5.25
6. F	5.30
7. B	5.30
8. H	5.30
9. B	5.30
10. G	5.17
11. C	5.6
12. H	5.36
13. C	5.5
14. J	5.24
15. A	5.3
16. F	5.22

17.   Answer: 13 stamps
Discussion
Tip: How can you use the calendar to find the number of stamps?

Samantha will need one stamp each evening from August 3 to August 15. She will not need a stamp for August 16 because she is leaving in the afternoon. Count the days on the calendar.

5.24 Identifies needed information and selects the steps necessary to solve multi-step word problems.

5.25 Solves one-, two-, and three-step word problems related to all appropriate fifth grade objectives including those presented orally and in writing; those in charts, tables, and graphs; and those with extraneous or insufficient information.

5.36 Adds, subtracts, multiplies and divides whole numbers up through four digits using both vertical and horizontal (rewrite vertically) presentation without and with regrouping (uses calculators for more laborious computations than these: four-digit addition and subtraction; multiplication and division of three- by two-digit numbers).

18.   Answer: Trevor, Martin, Jennifer

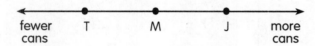

Discussion
Tip: How can drawing a diagram help you solve the problem?

Drawing a diagram may help you solve the problem. You can draw a number line and place dots for Trevor and Jennifer on the number line to show that Trevor collected fewer cans than Jennifer. Then read the next two sentences in the message to decide where to locate a point for Martin.

5.22 Compares and orders whole numbers, fractions, and decimals through hundredths.

5.24 Identifies needed information and selects the steps necessary to solve multi-step word problems.

5.26 Selects and uses appropriate strategies for solving problems (e.g., look for a pattern, guess and check, make an organized list, simplify the problem, work backwards).

19.   Answer: About $50
Discussion
Tip: How will rounding help you solve the problem?

Round each price up to the nearest dollar.

$17 + $10 + $11 + $14 = $52

5.3 Use estimation strategies (such as front-end, clustering, rounding, or reference point) to predict computational results of whole numbers, fractions, mixed numbers, and decimals.

5.25 Solves one-, two-, and three-step word problems related to all appropriate fifth grade objectives including those presented orally and in writing; those in charts, tables, and graphs; and those with extraneous or insufficient information.

© Harcourt

20.   Answer: 26 years old

Discussion

Tip: How is the aunt's age related to the ages of the three children?

The aunt is 5 years older than the sum of the ages of the children. Alicia is 4 years old. Esteban is 2 x 4, or 8 years old. Juan is 8 – 2, or 6 years old. Add to find the sum of their ages

$$4 + 8 + 6 = 18 \text{ years}$$

Since their aunt's age is 8 more than this sum, their aunt is 26 years old.

5.24 Identifies needed information and selects the steps necessary to solve multi-step word problems.

5.25 Solves one-, two-, and three-step word problems related to all appropriate fifth grade objectives including those presented orally and in writing; those in charts, tables, and graphs; and those with extraneous or insufficient information.

21.   Answer: 60 nickels

Discussion

Tip: How can you use factors and multiples to help you solve the problem?

You know that the number of nickels must be a multiple of 4 and of 5 and of 6. Since the nickels can be stacked in equal stacks of 5, the number of nickels has to end in a 5 or a 0. Since no multiple of 4 or 6 ends in a 5, the number of nickels must end in 0. Then you can use guess and check to see which numbers that end in zero have a factor of four and of six. The least number is 60.

Or, you can list the multiples of 4, of 5, and of 6 and find the lowest common multiple.
multiples of 4:   4, 8, 12, 16, 20, 24, 28, 32, 36, 40, 44, 48, 52, 56, <u>60</u>, ...
multiples of 5:   5, 10, 15, 20, 25, 30, 35, 40, 45, 50, 55, <u>60</u>, ...
multiples of 6:   6, 12, 18, 24, 30, 36, 42, 48, 54, <u>60</u>, ...

With 60 nickels Julia could make 15 stacks of four nickels each, 12 stacks of five nickels each, or 10 stacks of six nickels each.

5.19 Identifies factors and multiples of a given number, including prime factorization.

5.25 Solves one-, two-, and three-step word problems related to all appropriate fifth grade objectives including those presented orally and in writing; those in charts, tables, and graphs; and those with extraneous or insufficient information.

22.   Answer: The least number is 1,005,789.

Discussion

Tip: What do you know about place value?

Since this number has 7 digits, there will be a digit in every place through the the millions place.

To make the least number, put a 1 in the millions place, and zeros in the next two places. Then arrange the other digits so the highest digits have the lowest place value. You will end up with 9 in the ones place.

5.18 Identifies place value for whole numbers through millions, and decimals through hundredths. Determines the effect that changing a digit will have on the value of the number.

5.24 Identifies needed information and selects the steps necessary to solve multi-step word problems.

5.25 Solves one-, two-, and three-step word problems related to all appropriate fifth grade objectives including those presented orally and in writing; those in charts, tables, and graphs; and those with extraneous or insufficient information.

23. Answer: 4, 5, 6, 7, 8, or 9.

Discussion

Tip: How can you model the riddle on a number line?

You can use dots to show the numbers that match the first clue, and squares to show the numbers that match the second clue. The answer is the set of numbers that matches both clues.

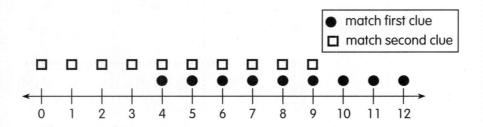

● match first clue
□ match second clue

5.22 Compares and orders whole numbers, fractions, and decimals through hundredths.

5.25 Solves one-, two-, and three-step word problems related to all appropriate fifth grade objectives including those presented orally and in writing; those in charts, tables, and graphs; and those with extraneous or insufficient information.

24. Answer: 58 students

Discussion

Tip: How do you subtract across zeros?

To find the difference, subtract the number of students in the fourth grade from the number of students in the fifth grade.

$$300 - 242 = 58 \text{ students}$$

There are 58 more students in the fifth grade.

Check: 58 + 242 = 300. The problem checks.

5.25 Solves one-, two-, and three-step word problems related to all appropriate fifth grade objectives including those presented orally and in writing; those in charts, tables, and graphs; and those with extraneous or insufficient information.

5.36 Adds, subtracts, multiplies and divides whole numbers up through four digits using both vertical and horizontal (rewrite vertically) presentation without and with regrouping (uses calculators for more laborious computations than these: four-digit addition and subtraction; multiplication and division of three- by two-digit numbers).

25. Answer: See discussion below.

Discussion

Tip: How many cans were collected in all? How many bottles were collected in all?

First find the total number of cans and the total number of bottles.

Cans: $45 + 82 + 133 = 260$

Bottles: $22 + 91 + 71 = 192$

Multiply these totals by the amount paid for each item to find the income from each.

Income from cans (3 cents per can):

$260 \times \$0.03 = \$7.80$

Income from bottles (5 cents per bottle):

$192 \times \$0.05 = \$9.60$

So, the students earned more income from recycling bottles.

5.24 Identifies needed information and selects the steps necessary to solve multi-step word problems.

5.30 Interprets and draws conclusions from charts, tables, and graphs (e.g., pictographs, and circle graphs).

5.36 Adds, subtracts, multiplies and divides whole numbers up through four digits using both vertical and horizontal (rewrite vertically) presentation without and with regrouping (uses calculators for more laborious computations than these: four-digit addition and subtraction; multiplication and division of three- by two-digit numbers).

26. Answer: See discussion below.

Discussion

Tip: What must you do to determine how many miles they will walk in three years?

Multiply for mail carrier.

$1{,}056 \times 3 = 3{,}168$ miles

Subtract to compare two professions.

Police officer	1,632
Mail carrier	− 1,056
difference	576 miles

There are 12 months in a year. Divide by 12 to find the number of miles walked each month:

$840 \div 12 = 70$

5.26 Selects and uses appropriate strategies for solving problems (e.g., look for a pattern, guess and check, make an organized list, simplify the problem, work backwards).

5.30 Interprets and draws conclusions from charts, tables, and graphs (e.g., pictographs, and circle graphs).

5.36 Adds, subtracts, multiplies and divides whole numbers up through four digits using both vertical and horizontal (rewrite vertically) presentation without and with regrouping (uses calculators for more laborious computations than these: four-digit addition and subtraction; multiplication and division of three- by two-digit numbers).

Georgia Quality Core Curriculum Objectives

5.2 Uses mental computation strategies such as counting up, counting back, compensation, compatible numbers and multiples of ten, hundred, or thousand, with whole numbers, fractions, and decimals, including money.

5.3 Use estimation strategies (such as front-end, clustering, rounding, or reference point) to predict computational results of whole numbers, fractions, mixed numbers, and decimals.

5.15 Develops procedures to determine and compute perimeter, area, and volume of various geometric figures through concrete experiences with covering, filling, and counting.

5.23 Solves simple problems requiring recall of basic facts.

5.25 Solves one-, two-, and three-step word problems related to all appropriate fifth grade objectives including those presented orally and in writing; those in charts, tables, and graphs; and those with extraneous or insufficient information.

5.26 Selects and uses appropriate strategies for solving problems (e.g., look for a pattern, guess and check, make an organized list, simplify the problem, work backwards).

5.35 Uses commutative, associative, and identity properties of addition and multiplication, and the distributive property of multiplication over addition.

5.36 Adds, subtracts, multiplies and divides whole numbers up through four digits using both vertical and horizontal (rewrite vertically) presentation without and with regrouping (uses calculators for more laborious computations than these: four-digit addition and subtraction; multiplication and division of three- by two-digit numbers).

1. Answer: E NOT HERE

Discussion

Tip: If you solve the problem and don't see your solution listed, mark NOT HERE as the answer.

Students should begin each problem by trying to find the solution. If they can solve the problem and don't see the answer listed, then they can eliminate the first four answer choices. If they have trouble solving the problem, encourage them to test each answer choice to see whether or not it can be eliminated.

Item Numbers	Georgia QCC Objectives
1. A	5.35
2. G	5.23
3. A	5.36
4. H	5.15
5. E	5.15
6. J	5.36
7. B	5.25
8. H	5.3
9. B	5.36
10. H	5.15
11. E	5.36
12. G	5.15
13. D	5.36
14. A	5.36
15. B	5.25
16. G	5.26
17. D	5.25
18. G	5.2
19. B	5.3
20. H	5.36
21. C	5.25
22. G	5.26

23. Answer: about 1,000 meals

Discussion

Tip: What number will you use to estimate the number of weeks in a year?

To find the number of meals you eat in one week, multiply the number of meals you eat in one day by the number of days in one week. 3 x 7 = 21. There are 52 weeks in one year. To estimate the number of meals you eat in one year, round 21 meals per week to 20 meals per week. Round 52 weeks per year to 50 weeks per year. Then use mental math to multiply:

> 20 x 50 = 1,000 so you eat about 1,000 meals in one year.

To estimate the number of meals you eat in ten years, multiply:

> 10 x 1,000 = 10,000 so you eat about 10,000 meals in ten years.

5.3 Use estimation strategies (such as front-end, clustering, rounding, or reference point) to predict computational results of whole numbers, fractions, mixed numbers, and decimals.

24. Answer: about $76 each month

Discussion

Tip: How much can she earn in one week?

Use addition and multiplication to find how much Anna can earn in 1 week. She can earn $19 in one week.

	allowance		dog		dishes		TOTAL
1 week	$ 5	+	$ 7	+	$ 7	=	$19
4 weeks	$20	+	$28	+	$28	=	$76

Then multiply the amount she can earn in one week by the number of weeks in a month.

> $19 × 4 = $76

So Anna can earn about $76 each month if she does every chore.

5.25 Solves one-, two-, and three-step word problems related to all appropriate fifth grade objectives including those presented orally and in writing; those in charts, tables, and graphs; and those with extraneous or insufficient information.

5.36 Adds, subtracts, multiplies and divides whole numbers up through four digits using both vertical and horizontal (rewrite vertically) presentation without and with regrouping (uses calculators for more laborious computations than these: four-digit addition and subtraction; multiplication and division of three- by two-digit numbers).

25. Answer: No; see discussion below.

Discussion

Tip: What operation can you use to solve the problem?

You can use multiplication or division.

There are 6 passengers and an adult driver for each van, so there are 7 people in each van. In 30 vans there is room for 30 x 7, or 210 people. So, 30 vans is not enough for 250 people.

> 250 people ÷ 7 people per van = 36 vans (round up so no one is left behind).

Another possible strategy: You can divide the number of people by the number of people who can ride in one van to find the number of vans you need for 250 people.

> 250 ÷ 7 = 35 R5   So, you would need 36 vans for 250 people.

5.24 Identifies needed information and selects the steps necessary to solve multi-step word problems.

© Harcourt

26. Answer: See discussion below.

Discussion

Tip: What are the factors of 36?

36 = 1 x 36 , 2 x 18, 3 x 12, 4 x 9, 6 x 6

You can use these factors to make these seating arrangements.

2 rows, 18 seats; 3 rows, 12 seats; 4 rows, 9 seats; 6 rows, 6 seats; 9 rows, 4 seats; 12 rows, 3 seats; 18 rows, 2 seats.

5.19 Identifies factors and multiples of a given number, including prime factorization.

5.36 Adds, subtracts, multiplies and divides whole numbers up through four digits using both vertical and horizontal (rewrite vertically) presentation without and with regrouping (uses calculators for more laborious computations than these: four-digit addition and subtraction; multiplication and division of three- by two-digit numbers).

27. Answer: Frank's car goes 12 miles on 1 gallon of gas. Helen's car goes 20 miles on 1 gallon of gas.

Discussion

Tip: What operations can you use to solve the problem?

Use division to find the miles per gallon.

Frank gets 180 miles ÷ 15 gallons = 12 miles per gallon.

Helen gets 180 miles ÷ 9 gallons = 20 miles per gallon.

Use subtraction to compare their fuel economy. 20 − 12 = 8

Helen's car gets 8 miles per gallon MORE than Frank's car.

5.26 Selects and uses appropriate strategies for solving problems (e.g., look for a pattern, guess and check, make an organized list, simplify the problem, work backwards).

5.36 Adds, subtracts, multiplies and divides whole numbers up through four digits using both vertical and horizontal (rewrite vertically) presentation without and with regrouping (uses calculators for more laborious computations than these: four-digit addition and subtraction; multiplication and division of three- by two-digit numbers).

28. Answer: Shana can use Design A or Design C.

Discussion

Tip: What operation can you use to solve the problem?

You can divide the number of beads Shana has by the number of beads in each design. If the quotient has no remainder, the design can work.

240 ÷ 24 = 10

240 ÷ 36 = 6 R24

240 ÷ 16 = 15

5.24 Identifies needed information and selects the steps necessary to solve multi-step word problems.

5.36 Adds, subtracts, multiplies and divides whole numbers up through four digits using both vertical and horizontal (rewrite vertically) presentation without and with regrouping (uses calculators for more laborious computations than these: four-digit addition and subtraction; multiplication and division of three- by two-digit numbers).

29. Answer: 538 miles per hour

Discussion

Tip: What operation can you use to solve the problem?

You can use division. You know the total distance. You also know that he flew for 5 hours. So you can divide the distance by the total number of hours to find the distance per hour.

$$2,690 \text{ miles} \div 5 \text{ hours} = 538 \text{ miles per hour}$$

5.26 Selects and uses appropriate strategies for solving problems (e.g., look for a pattern, guess and check, make an organized list, simplify the problem, work backwards).

30. Answer: The officer walks about 33 miles per week.

Discussion

Tip: What operation can you use?

You can use division. Divide the number of miles he walks in a year by the number of weeks he works in a year to find the number of miles he walks each week.

$$1,650 \text{ miles} \div 50 \text{ weeks per year}$$

$$1,650 \div 50 = 33 \text{ mi}$$

He walks 33 miles a week.

5.24 Identifies needed information and selects the steps necessary to solve multi-step word problems.

5.36 Adds, subtracts, multiplies and divides whole numbers up through four digits using both vertical and horizontal (rewrite vertically) presentation without and with regrouping (uses calculators for more laborious computations than these: four-digit addition and subtraction; multiplication and division of three- by two-digit numbers).

31. Answer: 33 years old

Discussion

Tip: How can you compare Stan's age to Alex's age?

Make a chart to keep track of all the ages. Begin with Alex, because his age is given in the problem. Then write a number sentence and solve for the ages of the other people.

Name	Age
Alex	5 yrs
Stan	2 x 5 = 10 yrs
Ben	10 - 2 = 8 yrs
Dad	5+ 10 + 8 + 10 = 33 yrs

5.26 Selects and uses appropriate strategies for solving problems (e.g., look for a pattern, guess and check, make an organized list, simplify the problem, work backwards).

5.36 Adds, subtracts, multiplies and divides whole numbers up through four digits using both vertical and horizontal (rewrite vertically) presentation without and with regrouping (uses calculators for more laborious computations than these: four-digit addition and subtraction; multiplication and division of three- by two-digit numbers).

32. Answer: 6,000 seats per section; about 30,000 people attended the concert.

Discussion

Tip: What operation can you use to solve the problem? How many seats are in each section?

To find the number of seat per section, you divide the total number of seats by the number of sections:

48,000 total ÷ 8 sections = 6,000 people per section

Count 6,000 for every section that is full or almost full. Count 3,000 for each section that is half full. Count 0 for each section that is mostly empty.

(Students should estimate in the thousands. A section that is almost empty balances out a section that is mostly full.)

Section A	Section B	Section C	Section D	Section E	Section F	Section G	Section H
almost full	almost full	full	full	half full	mostly empty	mostly empty	half full
6,000	6,000	6,000	6,000	3,000	0	0	3,000

5.24 Identifies needed information and selects the steps necessary to solve multi-step word problems.

5.36 Adds, subtracts, multiplies and divides whole numbers up through four digits using both vertical and horizontal (rewrite vertically) presentation without and with regrouping (uses calculators for more laborious computations than these: four-digit addition and subtraction; multiplication and division of three- by two-digit numbers).

Georgia Quality Core Curriculum Objectives

5.7 Relates a fraction to a part of a whole, a part of a set, and a point on a number line; uses models to determine equivalent fractions. Uses fractions with denominators of 2, 3, 4, 5, 6, 10, 16, or 100.

5.24 Identifies needed information and selects the steps necessary to solve multi-step word problems.

5.30 Interprets and draws conclusions from charts, tables, and graphs (e.g., pictographs, and circle graphs).

5.31 Using clustering to explore the concept of mean, median, and mode of a set of data and calculates the arithmetic mean.

5.32 Determines probability of a given event through exploration (more likely, less likely, equally likely, likely, or not likely).

5.33 Collects and organizes data into tallies, charts, and tables; determines appropriate scale and constructs bar graphs.

1. Answer: C population changes over 5 years

Discussion

Tip: What can you learn from the title of the graph and the labels on the axes?

Students should recognize that the graph's title tells that it discusses population size over a period of years. The axes tell what the population is and when it is measured. You may want to point out to students that they can rule out answers based on the title and labels—even before they have read the data in the graph.

Item Numbers	Georgia QCC Objectives
1. C	5.30
2. G	5.7, 5.30
3. D	5.30
4. H	5.24
5. D	5.30
6. G	5.30
7. D	5.33
8. J	5.30
9. B	5.32
10. G	5.30
11. B	5.7, 5.30
12. H	5.31
13. A	5.31
14. F	5.32
15. C	5.30

16.   Answer: August

Discussion

Tip: What does each bar on the graph represent?

Each bar represents the number of students buying shoes during one month. The height of each bar tells the number of students who bought shoes during that month. In August, students bought 22 pairs of shoes. Here are some possible conclusions based on the graph:

> In August, about twice as many students bought shoes as in July and September combined.

> In August, 17 more students bought shoes than in July.

5.24 Identifies needed information and selects the steps necessary to solve multi-step word problems.

5.30 Interprets and draws conclusions from charts, tables, and graphs (e.g., pictographs, and circle graphs).

17.   Answer: Janet is right. See discussion below.

Discussion

Tip: How many people are represented in the survey?

To find the number of people represented in the survey, you add all the votes:

$1 + 8 + 12 + 4 + 3 = 28$ people

Twelve out of 28 voted for pizza. This is fewer than half the people surveyed. So Janet is right. Most people did not vote for pizza.

5.24 Identifies needed information and selects the steps necessary to solve multi-step word problems.

5.30 Interprets and draws conclusions from charts, tables, and graphs (e.g., pictographs, and circle graphs).

18.   Answer: It is more likely that a family will have 1 girl and 1 boy.

Discussion

Tip: Does any combination of possibilities appear more than once?

Having a boy and a girl occurs more often. There are four possible outcomes. Two outcomes have one girl and one boy. So the probability of having one girl and one boy is 2 out of 4, or $\frac{2}{4}$, or $\frac{1}{2}$. Of the other outcomes, two boys will occur $\frac{1}{4}$ of the time and two girls will occur $\frac{1}{4}$ of the time.

5.24 Identifies needed information and selects the steps necessary to solve multi-step word problems.

5.32 Determines probability of a given event through exploration (more likely, less likely, equally likely, likely, or not likely).

© Harcourt

19. Answer: 149 students

Discussion

Tip: How many votes were cast for each choice?

Each tally mark represents one student's vote. First count the number of tally marks for each choice. Then add these numbers.

$$54 + 9 + 64 + 22 = 149$$

5.30 Interprets and draws conclusions from charts, tables, and graphs (e.g., pictographs, and circle graphs).

5.36 Adds, subtracts, multiplies and divides whole numbers up through four digits using both vertical and horizontal (rewrite vertically) presentation without and with regrouping (uses calculators for more laborious computations than these: four-digit addition and subtraction; multiplication and division of three- by two-digit numbers).

20. Answer: See discussion below.

Discussion

Tip: How many numbers on the spinner are odd numbers?
How many number on the spinner are greater than 4?

Of the numbers on the spinner, five out of six numbers are odd. Four out of six are greater than 4.

Statement will vary. Possible answers:

It is certain that someone will spin a number less than 10.

It is impossible to spin a 10.

It is likely that Cheryl will lose the game.

5.24 Identifies needed information and selects the steps necessary to solve multi-step word problems.

5.32 Determines probability of a given event through exploration (more likely, less likely, equally likely, likely, or not likely).

21. Answer: 12 outfits

Discussion

Tip: How can making a tree diagram help you solve the problem?

You can make a list that shows the pants in one column and the shirts in another. Then show each combination. The list helps you make sure that you are not counting the same outfit twice.

Red pants with red, white, blue, or multicolor shirt; white pants with red, white, blue, or multicolor shirt; blue pants with red, white, blue, or multicolor shirt.

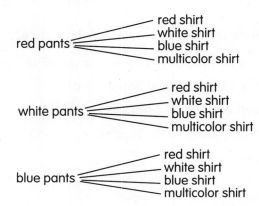

5.24 Identifies needed information and selects the steps necessary to solve multi-step word problems.

5.26 Selects and uses appropriate strategies for solving problems (e.g., look for a pattern, guess and check, make an organized list, simplify the problem, work backwards).

5.36 Adds, subtracts, multiplies and divides whole numbers up through four digits using both vertical and horizontal (rewrite vertically) presentation without and with regrouping (uses calculators for more laborious computations than these: four-digit addition and subtraction; multiplication and division of three- by two-digit numbers).

© Harcourt

22. Answer: See discussion below.

Discussion

Tip: How can you compare the data in the table to the data in the graph?

You can look at each line in the table and see if the number is shown accurately in the graph. Also, check the title, labels, and scale used.

Mistakes:

The title of the graph is different from the title of the table.

The vertical axis should show "Number of Marbles." The horizontal axis should be "Color."

Henry's scale in inconsistent. He omitted the number 2.

The number of blue marbles should show 5 in the graph.

The number of red marbles should show 3 in the graph.

5.26 Selects and uses appropriate strategies for solving problems (e.g., look for a pattern, guess and check, make an organized list, simplify the problem, work backwards).

5.30 Interprets and draws conclusions from charts, tables, and graphs (e.g., pictographs, and circle graphs).

23. Answer: $119

Discussion

Tip: What is the median of a set of data?

The median is the middle number in a set of data.

To find the median, arrange the numbers from low to high. If you have an odd number of items, choose the item in the center. If you have an even number of items, find the average (mean) of the two middle numbers.

From low to high, the prices are:

$60 $99 $119 $120 $155

Since $119 is the middle price, it is the median. Two pairs of skates cost less than $119, and two pairs cost more than $119.

5.31 Using clustering to explore the concept of mean, median, and mode of a set of data and calculates the arithmetic mean.

5.36 Adds, subtracts, multiplies and divides whole numbers up through four digits using both vertical and horizontal (rewrite vertically) presentation without and with regrouping (uses calculators for more laborious computations than these: four-digit addition and subtraction; multiplication and division of three- by two-digit numbers).

24. Answer: 8 possible outcomes

Discussion

Tip: How can you use a chart to help summarize the possible outcomes?

You can make a chart to show all the possible outcomes and check that you have found them all. The possible outcomes are listed below.

There are eight possible outcomes.

If Annabelle stays well, you will need only one tree diagram with 4 outcomes:

- All 3 are well.
- A and C are well, but B is sick.
- A and B are well, but C is sick.
- A is well, but B and C are sick.

5.24 Identifies needed information and selects the steps necessary to solve multi-step word problems.

5.26 Selects and uses appropriate strategies for solving problems (e.g., look for a pattern, guess and check, make an organized list, simplify the problem, work backwards).

Possible Outcomes

well	sick
A, B, C	---
A, B	C
A,C	B
A	B, C
B, C	A
B	A, C
---	A, B, C
C	A, B

Tree diagram: A is well → B is well → C is well / C is sick; B is sick → C is well / C is sick. A is sick → B is well → C is well / C is sick; B is sick → C is sick / C is well.

25. Answer: See discussion below.

Discussion

Tip: How can you change fractions to percents?

You can find an equivalent fraction with 100 in the denominator. The numerator in that fraction is the same as the percent.

Colin's Sports Card Collection

	Football	Baseball	Basketball	Hockey	TOTAL
Number of Boxes	4	32	12	2	50

Football: $\dfrac{4}{50} = 4 \times \dfrac{2}{50} \times 2 = \dfrac{8}{100}$, or 8%

Baseball : $\dfrac{32}{50} = 32 \times \dfrac{2}{50} \times 2 = \dfrac{64}{100}$, or 64%

Basketball: $\dfrac{12}{50} = 12 \times \dfrac{2}{50} \times 2 = \dfrac{24}{100}$, or 24%

Hockey: $\dfrac{2}{50} = 2 \times \dfrac{2}{50} \times 2 = \dfrac{4}{100}$, or 4%

5.24 Identifies needed information and selects the steps necessary to solve multi-step word problems.

5.33 Collects and organizes data into tallies, charts, and tables; determines appropriate scale and constructs bar graphs.

5.36 Adds, subtracts, multiplies and divides whole numbers up through four digits using both vertical and horizontal (rewrite vertically) presentation without and with regrouping (uses calculators for more laborious computations than these: four-digit addition and subtraction; multiplication and division of three- by two-digit numbers).

Georgia Quality Core Curriculum Objectives

5.1 Rounds whole numbers to nearest ten, hundred, or thousand; decimals to the nearest tenth or whole number; and fractions to the nearest whole number.

5.3 Use estimation strategies (such as front-end, clustering, rounding, or reference point) to predict computational results of whole numbers, fractions, mixed numbers, and decimals.

5.13 Selects appropriate customary and metric units of measure for length (including perimeter and circumference), area, capacity/volume, weight/mass, time, and temperature.

5.14 Uses customary and metric units to measure length, capacity/volume (use liquid and dry units), weight/mass, elapsed time and temperature (include measuring length to the nearest quarter inch, nearest millimeter and temperature below freezing).

5.16 Determines and estimates amounts of money.

5.20 Explores the concept of divisibility and develops rules for divisibility by 2, 3, 5, and 10.

5.21 Determines a pair of numbers or the missing element of a pair when given a relation or rule, and determines the relation or rule given pairs of numbers.

5.22 Compares and orders whole numbers, fractions, and decimals through hundredths.

5.24 Identifies needed information and selects the steps necessary to solve multi-step word problems.

5.25 Solves one-, two-, and three-step word problems related to all appropriate fifth grade objectives including those presented orally and in writing; those in charts, tables, and graphs; and those with extraneous or insufficient information.

5.26 Selects and uses appropriate strategies for solving problems (e.g., look for a pattern, guess and check, make an organized list, simplify the problem, work backwards).

5.27 Predicts measurement by using strategies such as walking off and rough comparison.

5.28 Given simulation, chooses the most appropriate method of computation (mental computation, paper and pencil, or calculator).

5.36 Adds, subtracts, multiplies and divides whole numbers up through four digits using both vertical and horizontal (rewrite vertically) presentation without and with regrouping (uses calculators for more laborious computations than these: four-digit addition and subtraction; multiplication and division of three- by two-digit numbers).

1. Answer: B $n = 0.66$

Discussion

Tip: Look for patterns in the second factor of the multiplication equations.

Students should begin by looking for a pattern in the numbers. The first factor is the same, but the second factor in each equation changes when multiplied by 0.1. Ask:

How does the second factor change in each equation?
How does this change influence the answer?
If the pattern continued, what would be the next equation?

Try solving other equations using similar patterns and discussing the patterns found.

Item Numbers	Georgia QCC Objectives
1. B	5.21, 5.26
2. H	5.28
3. A	
4. J	5.3
5. C	
6. G	5.1, 5.16
7. A	5.20, 5.24
8. H	5.21, 5.26
9. C	5.16, 5.25, 5.36
10. F	
11. C	5.14
12. H	5.13
13. C	
14. G	5.16, 5.25
15. B	5.16, 5.25
16. J	5.25
17. D	5.27
18. G	5.14, 5.22
19. B	5.25
20. J	5.25
21. B	
22. H	5.14

© Harcourt

23. Answer: See discussion below.

Discussion
Tip: What operation can you use to solve the problem?

You can use multiplication to find the cost for each multiple of hair scrunchies.

Charmaine's SuperFine Hair Scrunchies

Number	4	8	12	16	20
Cost	$3.56	$7.12	$10.68	$14.24	$17.80

5.21 Determines a pair of numbers or the missing element of a pair when given a relation or rule, and determines the relation or rule given pairs of numbers.

5.33 Collects and organizes data into tallies, charts, and tables; determines appropriate scale and constructs bar graphs.

5.36 Adds, subtracts, multiplies and divides whole numbers up through four digits using both vertical and horizontal (rewrite vertically) presentation without and with regrouping (uses calculators for more laborious computations than these: four-digit addition and subtraction; multiplication and division of three- by two-digit numbers).

24. Answer: two of these: feet, yards, or meters
Discussion
Tip: How will she measure a distance of five or ten steps?

Twelve inches (one standard foot) is about as long as Sheila's foot. A yard, or a meter, is about as long as one stride. All these units measure length or distance.

5.13 Selects appropriate customary and metric units of measure for length (including perimeter and circumference), area, capacity/volume, weight/mass, time, and temperature.

5.27 Predicts measurement by using strategies such as walking off and rough comparison.

25. Answer: millimeters

Discussion

Tip: Will you use smaller units or larger units of measure?

By using smaller units, you can measure the illustration more accurately. You won't need to use fractions of a unit. This hummingbird is 89 millimeters long.

5.13 Selects appropriate customary and metric units of measure for length (including perimeter and circumference), area, capacity/volume, weight/mass, time, and temperature.

26   Answer: 162, 486; Every number is 3 times larger than the preceding number.

Discussion

Tip: What operation can you use to find the next number using the number before?

You can multiply each number by three to find the next number.

    2 x 3   = 6
    6 x 3   = 18
    18 x 3   = 54
    54 x 3   = 162
    162 x 3 = 486

Students may try finding addition patterns first. Encourage them to look for the differences between consecutive numbers. When the differences are irregular, the pattern is more likely to be a multiplication pattern.

5.21 Determines a pair of numbers or the missing element of a pair when given a relation or rule, and determines the relation or rule given pairs of numbers.

5.24 Identifies needed information and selects the steps necessary to solve multi-step word problems.

5.26 Selects and uses appropriate strategies for solving problems (e.g., look for a pattern, guess and check, make an organized list, simplify the problem, work backwards).

27.   Answer: She cannot use all 19 plants to make equal rows with an equal number of plants in each row.

Discussion

Tip: Is 19 a prime or a composite number?

19 is a prime number. The factors of 19 are 1 and 19.

If Alana takes away one plant, she will have 18 plants. Then she can make 2 rows of 9, or 3 rows of 6, 6 rows of 3, or 9 rows of 2.

5.36 Adds, subtracts, multiplies and divides whole numbers up through four digits using both vertical and horizontal (rewrite vertically) presentation without and with regrouping (uses calculators for more laborious computations than these: four-digit addition and subtraction; multiplication and division of three- by two-digit numbers).

28. Answer: 1 d; 2 c; 3 b; 4 a

Discussion

Tip: What benchmark decimals and benchmark fractions can help you solve the problem? How can reading each number aloud help you decide?

You know that 0.5 or 0.50 is exactly half. You know that $\frac{10}{10}$ is exactly one whole.

Reading each number aloud helps you hear its relationship to a benchmark. Fifty-two hundredths is close to 50 hundredths which is a decimal for half. So 0.52 is a little more than half.

5.24 Identifies needed information and selects the steps necessary to solve multi-step word problems.

29. Answer: See discussion below.

Discussion

Tip: Compare the two measures of engine capacity for the Jaguar. What ratio do you see?

You can multiply horsepower by 10 to get cubic centimeters. To go from cubic centimeters to horsepower, divide by 10.

The rule: the number of cubic centimeters is 10 times the horsepower. You can move the decimal point in the number of cubic centimeters one place to the left to find the horsepower.

Engine Capacity

Name of Car	Engine Capacity (in cubic centimeters)	Engine Capacity (in horsepower)
lightest car	2.5	0.25
Mazda RX2	900	90
AMC Hornet	1,280	128
Volvo Sedan	1,780	178
Ford Thunderbird	3,000	300
Jaguar E-car	4,235	423.5

5.21 Determines a pair of numbers or the missing element of a pair when given a relation or rule, and determines the relation or rule given pairs of numbers.

5.33 Collects and organizes data into tallies, charts, and tables; determines appropriate scale and constructs bar graphs.

30. Answer: Mandy

Discussion:

Tip: What information do you need to find out how much money Mandy spent?

You need to know that there are 16 ounces in 1 pound, and 32 ounces in 2 pounds.

Sharon spent $2.95 x 2 = $5.90   Mandy spent $0.25 x 32 = $8.00

First multiply the price per pound times the number of pounds Sharon bought to find out how much Sharon spent. $2.95 x 2 = $5.90. Since there are 16 ounces in a pound, you can find out how many ounces of candy Mandy bought by multiplying 16 ounces x 2 = 32 ounces. She paid $0.25 per ounce, so you can multiply 32 x $0.25 = $8.00. Mandy spent $8.00. Therefore, Mandy spent more.

5.16 Determines and estimates amounts of money.

5.36 Adds, subtracts, multiplies and divides whole numbers up through four digits using both vertical and horizontal (rewrite vertically) presentation without and with regrouping (uses calculators for more laborious computations than these: four-digit addition and subtraction; multiplication and division of three- by two-digit numbers).

31. Answer: It will cost less to use 5-inch tiles.

Discussion

Tip: How does drawing a model help you solve the problem? How can drawing a diagram help solve the problem?

When you draw a diagram, you can easily see how many tiles go into the 15-inch sample square. Count how many 5-inch tiles it would take to cover one square. Count how many 3-inch tiles it would take to cover the other square.

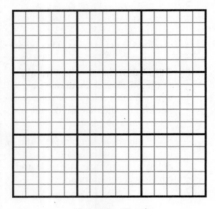

5-inch tiles
9 tiles needed

3-inch tiles
25 tiles needed

Then, multiply each number by the cost per tile.

For 5-inch tiles: 9 tiles × $1.76 = $15.84

For 3-inch tiles: 25 tiles × $0.67 = $16.75

So, it will cost less to use 5-inch tiles.

5.15 Develops procedures to determine and compute perimeter, area, and volume of various geometric figures through concrete experiences with covering, filling,

5.24 Identifies needed information and selects the steps necessary to solve multi-step word problems.

5.26 Selects and uses appropriate strategies for solving problems (e.g., look for a pattern, guess and check, make an organized list, simplify the problem, work backwards).and counting.

© Harcourt

32. Answer: See discussion below.

Discussion

Tip: What will each person be served at the picnic? How can rounding help you solve the problem?

Each person will be served two hotdogs, two rolls and two cans of juice.

Brenda's shopping list may look like this:

Brenda's Shopping List

Food	Number Needed	Number in Package	Number of Packages Needed		
Hotdogs	96	10	10		
Rolls	96	8	12		
Juice	96	24	4		

To estimate her budget, round the item prices. Add the rounded item price in the table:

Brenda's Shopping List

Food	Number Needed	Number in Package	Number of Packages Needed	Rounded Price per Package	Cost for This Item
Hotdogs	96	10	10	$2.00	$20.00
Rolls	96	8	12	$1.00	$12.00
Juice	96	24	4	$4.00	$16.00
				Total Cost:	**$48.00**

She can stay within her budget.

5.3 Use estimation strategies (such as front-end, clustering, rounding, or reference point) to predict computational results of whole numbers, fractions, mixed numbers, and decimals.

5.33 Collects and organizes data into tallies, charts, and tables; determines appropriate scale and constructs bar graphs.

5.36 Adds, subtracts, multiplies and divides whole numbers up through four digits using both vertical and horizontal (rewrite vertically) presentation without and with regrouping (uses calculators for more laborious computations than these: four-digit addition and subtraction; multiplication and division of three- by two-digit numbers).

Georgia Quality Core Curriculum Objectives

5.4 Adds, subtracts, and multiplies fractions and mixed numbers with like and unlike denominators, using vertical and horizontal presentation.

5.7 Relates a fraction to a part of a whole, a part of a set, and a point on a number line; uses models to determine equivalent fractions. Uses fractions with denominators of 2, 3, 4, 5, 6, 10, 16, or 100.

5.19 Identifies factors and multiples of a given number, including prime factorization.

5.22 Compares and orders whole numbers, fractions, and decimals through hundredths.

5.25 Solves one-, two-, and three-step word problems related to all appropriate fifth grade objectives including those presented orally and in writing; those in charts, tables, and graphs; and those with extraneous or insufficient information.

5.26 Selects and uses appropriate strategies for solving problems (e.g., look for a pattern, guess and check, make an organized list, simplify the problem, work backwards).

5.34 Identifies the number or symbol (+, −, x, —, <, >, =) that makes a given number sentence true.

1. Answer: A $2\frac{2}{3}$

Discussion

Tip: What two steps are needed to write a fraction as a mixed number?

Students should recognize that you have to first divide the numerator by the denominator and then write the remainder as a fraction in simplest form. Show them how to divide the numerator by the denominator. Then ask:

> How would you write the remainder as a fraction in simplest form? What does it mean to be in simplest form?

Item Numbers	Georgia QCC Objectives
1. A	5.7
2. H	5.22
3. A	5.22, 5.34
4. H	5.7
5. C	5.19
6. J	5.7
7. A	5.7
8. J	5.7, 5.25, 5.26
9. B	5.4
10. J	5.4, 5.25
11. A	5.4
12. J	5.4, 5.25
13. D	5.4
14. J	5.4, 5.25
15. C	5.4
16. H	5.4, 5.25
17. D	5.7
18. F	5.4, 5.25

19. Answer: one fourth $\left(\frac{1}{4}\right)$

Discussion

Tip: How can you use equivalent fractions to solve the problem?

You can use twelfths to form equivalent fractions. Then you can add the parts eaten to find out how much of the pie was eaten by the judges.

One fourth, one third and one sixth were eaten.

$$\frac{1}{4} + \frac{1}{3} + \frac{1}{6} = \frac{3}{12} + \frac{4}{12} + \frac{2}{12} = \frac{9}{12} = \frac{3}{4}$$

Three fourths was eaten. So, one fourth of the pie was left.

5.7 Relates a fraction to a part of a whole, a part of a set, and a point on a number line; uses models to determine equivalent fractions. Uses fractions with denominators of 2, 3, 4, 5, 6, 10, 16, or 100.

20. Answer: They ate the same amount.

Discussion

Tip: How can you use equivalent fractions to solve the problem?

To find the solution, you could find common denominators to compare fractions.

$$\frac{1}{3} = \frac{2}{6}$$
$$\frac{2}{6} = \frac{2}{6}$$

This shows they ate exactly the same amount of apple pie. The fractions are equivalent. You could also draw a picture or use fraction bars.

5.7 Relates a fraction to a part of a whole, a part of a set, and a point on a number line; uses models to determine equivalent fractions. Uses fractions with denominators of 2, 3, 4, 5, 6, 10, 16, or 100.

5.22 Compares and orders whole numbers, fractions, and decimals through hundredths.

21. Answers: See discussion below.

Discussion

Tip: What fraction of all people do left-handed people represent?

Left-handed people represent $\frac{1}{10}$ of all people. Each group described can be divided by 10 to get the number in one tenth of the total.

Mr. Kant should buy about $100 \div 10$, or 10 left-handed scissors.

The rule: since left-handers comprise one tenth of the population, you should find about one tenth of all the people in a group to find the number of left-handed people. You can divide by ten to find one tenth of any whole number.

5.4 Adds, subtracts, and multiplies fractions and mixed numbers with like and unlike denominators, using vertical and horizontal presentation.

5.36 Adds, subtracts, multiplies and divides whole numbers up through four digits using both vertical and horizontal (rewrite vertically) presentation without and with regrouping (uses calculators for more laborious computations than these: four-digit addition and subtraction; multiplication and division of three- by two-digit numbers).

22. Answer: There are 6 yellow balloons.

Discussion

Tip: How do you find a fraction of a number?

There are many different ways to solve this problem. One way is to add $\frac{1}{5}$ and $\frac{1}{2}$, which equals $\frac{7}{10}$. Find $\frac{7}{10}$ of 20, or 14. Subtract 14 from 20. This leaves 6 green balloons. You could also find $\frac{1}{5}$ of 20, or 4, and $\frac{1}{2}$ of 20, or 10. Add 4 and 10, subtract this number from 20. This also leaves 6 green balloons.

5.4 Adds, subtracts, and multiplies fractions and mixed numbers with like and unlike denominators, using vertical and horizontal presentation.

5.36 Adds, subtracts, multiplies and divides whole numbers up through four digits using both vertical and horizontal (rewrite vertically) presentation without and with regrouping (uses calculators for more laborious computations than these: four-digit addition and subtraction; multiplication and division of three- by two-digit numbers).

© Harcourt

23. Answer: greastest: cheese, least: pepperoni
Discussion

Tip: How can you order mixed numbers?

You can use a number line to order the mixed numbers.

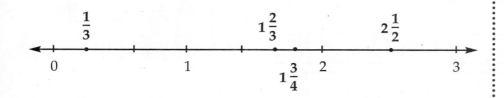

Looking at the number line, Darrell needs more cups of cheese than any other ingredient. Darrell needs the least amount of pepperoni.

5.22 Compares and orders whole numbers, fractions, and decimals through hundredths.

5.34 Identifies the number or symbol (+, −, x, —, <, >, =) that makes a given number sentence true.

24. Answer: Food
Discussion

Tip: How can you compare fractions?

To compare fractions, find a common denominator.

Change $\frac{3}{8}$ to $\frac{6}{16}$ so that you can compare it with $\frac{5}{16}$. When two

fractions have common denominators, you can look at the number

in the numerator. The number 6 is larger than the number 5, so

$\frac{6}{16} > \frac{5}{16}$.

Clayton spends more money on food.

5.7 Relates a fraction to a part of a whole, a part of a set, and a point on a number line; uses models to determine equivalent fractions. Uses fractions with denominators of 2, 3, 4, 5, 6, 10, 16, or 100.

5.22 Compares and orders whole numbers, fractions, and decimals through hundredths.

5.34 Identifies the number or symbol (+, −, x, —, <, >, =) that makes a given number sentence true.

5.4 Adds, subtracts, and multiplies fractions and mixed numbers with like and unlike denominators, using vertical and horizontal presentation.

5.26 Selects and uses appropriate strategies for solving problems (e.g., look for a pattern, guess and check, make an organized list, simplify the problem, work backwards).

25. Answer: $4\frac{1}{2}$ mi

Discussion

Tip: How will making a model help you add fractions?

Draw a model of fraction bars to help you.

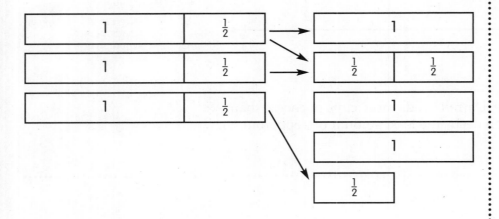

Kevin walked $4\frac{1}{2}$ miles this week.

5.4 Adds, subtracts, and multiplies fractions and mixed numbers with like and unlike denominators, using vertical and horizontal presentation.

26. Answer: $\frac{5}{12}$ in.

Discussion

Tip: How do you subtract fractions that have unlike denominators?

To subtract unlike fractions, use the LCD to change the fractions to like fractions. Since the LCD of $\frac{3}{4}$ and $\frac{1}{3}$ is twelfths use the LCD to write equivalent fractions and then subtract.

$$5\frac{3}{4} - 5\frac{1}{3} = 5\frac{9}{12} - 5\frac{4}{12} = \frac{5}{12}$$

After two weeks, $\frac{5}{12}$ in. of Jennica's pencil was gone.

27. Answer: See discussion below.

Discussion

Tip: What is the number of hours in Lisa's school day? What can you see clearly from the circle graph?

Lisa's school day is 8 hours long.

Lisa's School Day

Activity	Time Spent	Fraction of School Day	Fraction in Simplest Form
Recess	1 hour	$\frac{1}{8}$	$\frac{1}{8}$
Class Learning Time	5 hours	$\frac{5}{8}$	$\frac{5}{8}$
Homework	2 hours	$\frac{2}{8}$	$\frac{1}{4}$

The circle graph shows how Lisa spends her time during an 8-hour school day.

Students will make many valid observations. Here are some possible observations. You can tell from the graph that she spends most of her time in class and the least amount of time at recess. She spends more time doing homework than at recess. She spends less time doing homework than in class learning.

Lisa's School Day

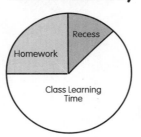

5.4 Adds, subtracts, and multiplies fractions and mixed numbers with like and unlike denominators, using vertical and horizontal presentation.

5.21 Determines a pair of numbers or the missing element of a pair when given a relation or rule, and determines the relation or rule given pairs of numbers.

5.33 Collects and organizes data into tallies, charts, and tables; determines appropriate scale and constructs bar graphs.

28. Answer: See discussion below.

Discussion

Tip: How many books were donated altogether? What fraction of the circle graph does each book type represent?

Thirty books were donated altogether. Sports take up three tenths of the circle graph; mystery and adventure each take up two tenths of the graph, and humor, history, and science each show one tenth of the graph.

See chart and circle graph below.

Library Book Donations

Type of Book	Number of Books Donated	Fraction of Books Donated	
Sports	9	$\frac{9}{30}$	$\frac{3}{10}$
Humor	3	$\frac{3}{30}$	$\frac{1}{10}$
Mystery	6	$\frac{6}{30}$	$\frac{2}{10}$
Adventure	6	$\frac{6}{30}$	$\frac{2}{10}$
History	3	$\frac{3}{30}$	$\frac{1}{10}$
Science	3	$\frac{3}{30}$	$\frac{1}{10}$

Library Book Donations

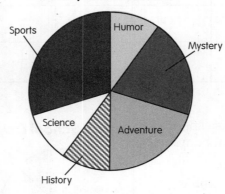

Sample statements:

Sports was the most popular category.

Mystery and adventure were the next most popular.

One tenth of the students donated humor books.

5.26 Selects and uses appropriate strategies for solving problems (e.g., look for a pattern, guess and check, make an organized list, simplify the problem, work backwards).

5.33 Collects and organizes data into tallies, charts, and tables; determines appropriate scale and constructs bar graphs.

Georgia Quality Core Curriculum Objectives

5.3 Use estimation strategies (such as front-end, clustering, rounding, or reference point) to predict computational results of whole numbers, fractions, mixed numbers, and decimals.

5.4 Adds, subtracts, and multiplies fractions and mixed numbers with like and unlike denominators, using vertical and horizontal presentation.

5.14 Uses customary and metric units to measure length, capacity/volume (use liquid and dry units), weight/mass, elapsed time and temperature (include measuring length to the nearest quarter inch, nearest millimeter and temperature below freezing).

5.22 Compares and orders whole numbers, fractions, and decimals through hundredths.

5.25 Solves one-, two-, and three-step word problems related to all appropriate fifth grade objectives including those presented orally and in writing; those in charts, tables, and graphs; and those with extraneous or insufficient information.

1. Answer: D $n = 1\frac{1}{2}$

Discussion

Tip: How do you add fractions with like denominators?

Students should understand that to add fractions with like denominators they simply add the numerators. Then they place the sum over the like denominator.

Remind students to simplify the answer if they can. You might want to use models to show addition and subtraction of fractions with like denominators.

Item Numbers	Georgia QCC Objectives
1. D	5.4
2. F	5.4
3. B	5.25
4. H	5.4
5. B	5.3
6. H	5.4
7. B	5.14, 5.25
8. H	5.4
9. C	5.25
10. G	5.22
11. D	5.4
12. J	5.4
13. C	5.14
14. J	5.25
15. C	5.25
16. F	5.14

17. Answer: 4 blue, 3 yellow, 5 green

Discussion

Tip: How can you use equivalent fractions to solve the problem?

Use 12 as a common denominator and write equivalent fractions.

BLUE: $\frac{1}{3}$ dozen $\times \frac{4}{4} = \frac{4}{12}$ (four out of twelve are blue)

YELLOW: $\frac{1}{4}$ dozen $\times \frac{3}{3} = \frac{3}{12}$ (three out of twelve are yellow)

GREEN: $\frac{5}{12}$ (five out of twelve are green)

If students are not ready to work with equivalent fractions, have them color one third of the eggs blue, one fourth of the eggs yellow, and five-twelfths of the eggs green.

5.4 Adds, subtracts, and multiplies fractions and mixed numbers with like and unlike denominators, using vertical and horizontal presentation.

5.7 Relates a fraction to a part of a whole, a part of a set, and a point on a number line; uses models to determine equivalent fractions. Uses fractions with denominators of 2, 3, 4, 5, 6, 10, 16, or 100.

18. Answer: 160 servings

Discussion

Tip: How many cups are in one gallon?

4 cups = 1 quart

4 quarts = 1 gallon

1 gallon = 4 x 4 = 16 cups

10 gallons = 10 x 16 = 160 cups

5.14 Uses customary and metric units to measure length, capacity/volume (use liquid and dry units), weight/mass, elapsed time and temperature (include measuring length to the nearest quarter inch, nearest millimeter and temperature below freezing).

19. Answer: harmonica and/or book, OR camera

Discussion

Tip: There are 16 ounces in 1 pound. How will you use this to find an answer?

Convert the weight limit to ounces. 1 pound = 16 ounces

16 ounces – 3 ounces for yo-yo = 13 ounces remaining for other items.

5.14 Uses customary and metric units to measure length, capacity/volume (use liquid and dry units), weight/mass, elapsed time and temperature (include measuring length to the nearest quarter inch, nearest millimeter and temperature below freezing).

20.   Answer: $14\frac{1}{2}$ hours

Discussion

Tip: There are 60 minutes in 1 hour. What operation can you use to convert from minutes to hours?

To convert from a smaller unit to a larger unit, you divide. There are 60 minutes in 1 hour.

$$870 \div 60 = 14 \text{ r}30$$

Since there are 30 minutes in a half-hour, you can interpret the remainder.

$$870 \div 60 = 14.5 \text{ or } 14\frac{1}{2} \text{ hours}$$

5.14 Uses customary and metric units to measure length, capacity/volume (use liquid and dry units), weight/mass, elapsed time and temperature (include measuring length to the nearest quarter inch, nearest millimeter and temperature below freezing).

5.36 Adds, subtracts, multiplies and divides whole numbers up through four digits using both vertical and horizontal (rewrite vertically) presentation without and with regrouping (uses calculators for more laborious computations than these: four-digit addition and subtraction; multiplication and division of three- by two-digit numbers).

21.   Answer: 600 meters

Discussion

Tip: How can drawing a model help you solve the problem?

A model can show what part of the iceberg is below the surface and what part is above the water.

If $\frac{3}{4}$ of the iceberg is below the surface, $\frac{1}{4}$ is above the surface. The part below the surface is three times as high as the part visible above the water.

$$150 \text{ m} \times 4 = 600 \text{ m}$$

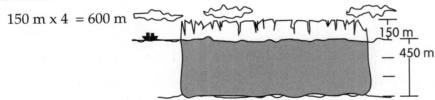

150 m

450 m

5.4 Adds, subtracts, and multiplies fractions and mixed numbers with like and unlike denominators, using vertical and horizontal presentation.

5.36 Adds, subtracts, multiplies and divides whole numbers up through four digits using both vertical and horizontal (rewrite vertically) presentation without and with regrouping (uses calculators for more laborious computations than these: four-digit addition and subtraction; multiplication and division of three- by two-digit numbers).

22. Answer: No. There is not enough wood.

Discussion

Tip: How do you add mixed numbers?

First, add $5\frac{1}{2}$ and $1\frac{1}{3}$. To add mixed numbers, use the LCD to change the fractions to like fractions. Since the LCD of $\frac{1}{2}$ and $\frac{1}{3}$ is sixths, use the LCD to write equivalent fractions. Add the fractions. Then add the whole numbers.

$$5\frac{1}{2} + 1\frac{1}{3} = 5\frac{3}{6} + 1\frac{2}{6} = 6\frac{5}{6}$$

Since $6\frac{5}{6} < 7$, Jerome and Clara do not have enough wood.

5.4 Adds, subtracts, and multiplies fractions and mixed numbers with like and unlike denominators, using vertical and horizontal presentation.

5.7 Relates a fraction to a part of a whole, a part of a set, and a point on a number line; uses models to determine equivalent fractions. Uses fractions with denominators of 2, 3, 4, 5, 6, 10, 16, or 100.

5.22 Compares and orders whole numbers, fractions, and decimals through hundredths.

23. Answer: Juice glass

Discussion

Tip: Which container has a capacity that is closest to 8 ounces?

A gallon jug or a mixing bowl would be far too big. A teaspoon is far too small. The juice glass probably holds 7–9 ounces. Sharifa can adjust the amount of flour she puts in by estimating the size of the glass.

5.13 Selects appropriate customary and metric units of measure for length (including perimeter and circumference), area, capacity/volume, weight/mass, time, and temperature.

24. Answer: 4 in.

Discussion

Tip: How can you change inches to feet?

Use the fact that 12 in. = 1 ft. Change 50 in. to feet by dividing 50 by 12. The remainder, 2, means 2 inches.

$$50 \text{ in.} = 4 \text{ ft } 2 \text{ in.}$$

Then find how many inches are in $\frac{1}{2}$ ft.

$\frac{1}{2}$ ft. $= \frac{1}{2} \times 12 = 6$ in.

To ride the roller coaster, he needs to be 4 ft 6 in. Chris needs to grow 4 more inches.

5.4 Adds, subtracts, and multiplies fractions and mixed numbers with like and unlike denominators, using vertical and horizontal presentation.

5.14 Uses customary and metric units to measure length, capacity/volume (use liquid and dry units), weight/mass, elapsed time and temperature (include measuring length to the nearest quarter inch, nearest millimeter and temperature below freezing).

© Harcourt

25. Answer: See discussion below.

Discussion

Tip: How can you use the diagram to help solve the problem?

You can tell by looking at the diagram that you need 48 inches of 60-inch-wide fabric. Since there are 36 inches in 1 yard, divide 48 by 36 to find the number of yards of fabric you need.

$$48 \div 36 = 1\frac{1}{3}$$

The gold paint covers 12 inches x 12 inches, or 144 square inches, or 1 square foot.

Drawings will vary. The gold should cover no more than 144 square inches in the cape. Students also might notice that each square on the grid represents 1 square foot so the total gold area should be less than or equal to one square on the grid. The cape is 30 square feet, so the gold should be $\frac{1}{30}$ of the total, or less.

Students may explain that they measured or estimated the number of square inches in their gold designs.

5.14 Uses customary and metric units to measure length, capacity/volume (use liquid and dry units), weight/mass, elapsed time and temperature (include measuring length to the nearest quarter inch, nearest millimeter and temperature below freezing).

5.15 Develops procedures to determine and compute perimeter, area, and volume of various geometric figures through concrete experiences with covering, filling, and counting.

5.24 Identifies needed information and selects the steps necessary to solve multi-step word problems.

26. Answer: See discussion below.

Discussion

Tip: What is the difference in the capacity of the two glasses?

Scott can measure 2 ounces by filling the 7-ounce glass and pouring 5 ounces into the other glass.

Possible answers:

7 oz + 7 oz + 2 oz = 16 oz

7 oz + 5 oz + 2 oz + 2 oz = 16 oz

5.14 Uses customary and metric units to measure length, capacity/volume (use liquid and dry units), weight/mass, elapsed time and temperature (include measuring length to the nearest quarter inch, nearest millimeter and temperature below freezing).

5.24 Identifies needed information and selects the steps necessary to solve multi-step word problems.

© Harcourt

Georgia Quality Core Curriculum Objectives

5.9 Identifies and distinguishes among point, ray, line, line segment, and angle.

5.10 Determines line of symmetry and identifies geometric relations (e.g., parallel to, perpendicular to, intersect, horizontal, vertical, similar, congruent, flips, slides, and turns).

5.11 Makes models of plane and solid figures and sorts and classifies these models according to characteristics such as number of sides, angles, vertices, faces, edges, tessellations, and lines of symmetry, (include triangles, quadrilaterals, polygons, circles, cones, cylinders, rectangular prisms, and pyramids).

5.12 Uses ordered pairs of numbers to locate points on a grid or map and determine the ordered pair for a given point.

5.15 Develops procedures to determine and compute perimeter, area, and volume of various geometric figures through concrete experiences with covering, filling, and counting.

5.25 Solves one-, two-, and three-step word problems related to all appropriate fifth grade objectives including those presented orally and in writing; those in charts, tables, and graphs; and those with extraneous or insufficient information.

1. Answer: D $\overline{AC}$

Discussion

Tip: Think about what the word "parallel" means. Draw a picture if that would help you.

Students should recognize that parallel lines do not intersect and are the same distance apart at all points.

Help students solve the problem by looking closely at the figure. Ask:

> Where is $\overline{CD}$ located?
> What line segments are part of lines that don't intersect $\overline{CD}$?
> Are any of these line segments listed as possible answers?

Item Numbers	Georgia QCC Objectives
1. A	5.10
2. F	5.10
3. D	5.9
4. J	5.9
5. B	5.15
6. F	5.10
7. D	5.10
8. H	5.12
9. A	5.11
10. G	5.10
11. D	5.11
12. F	5.25
13. D	5.25
14. J	5.15
15. C	
16. F	
17. B	5.25
18. H	

19.   Answer: See discussion below.
Discussion
Tip: What makes each type of polygon unique?

The number of sides and angles lets you tell the difference between different polygons.

Students should recognize triangles, squares and rectangles in the drawing.

Congruent figures are the same size and shape. Squares are the congruent polygons in this drawing. Each window is a medium-sized square surrounding four smaller congruent squares.

5.10 Determines line of symmetry and identifies geometric relations (e.g., parallel to, perpendicular to, intersect, horizontal, vertical, similar, congruent, flips, slides, and turns).

5.11 Makes models of plane and solid figures and sorts and classifies these models according to characteristics such as number of sides, angles, vertices, faces, edges, tessellations, and lines of symmetry, (include triangles, quadrilaterals, polygons, circles, cones, cylinders, rectangular prisms, and pyramids).

20.   Answer: See discussion below.
Discussion
Tip: How do you identify points for ordered pairs in a grid?

The first number tells how far to move right from 0. The second number tells how far to move up from 0. For example, to locate point *A*, begin at 0, move 1 space to the right and 2 spaces up.

The finished drawing is a polygon with four sides, or quadrilateral.

5.11 Makes models of plane and solid figures and sorts and classifies these models according to characteristics such as number of sides, angles, vertices, faces, edges, tessellations, and lines of symmetry, (include triangles, quadrilaterals, polygons, circles, cones, cylinders, rectangular prisms, and pyramids).

5.12 Uses ordered pairs of numbers to locate points on a grid or map and determine the ordered pair for a given point.

21.   Answer: The figure is a parallelogram. (Accept any of these: square, rectangle, parallelogram, rhombus)
Discussion
Tip: What math words will help you understand the clues?

You could have drawn a square, rectangle, parallelogram, or rhombus. These are all quadrilateral polygons with pairs of parallel sides.

5.10 Determines line of symmetry and identifies geometric relations (e.g., parallel to, perpendicular to, intersect, horizontal, vertical, similar, congruent, flips, slides, and turns).

5.11 Makes models of plane and solid figures and sorts and classifies these models according to characteristics such as number of sides, angles, vertices, faces, edges, tessellations, and lines of symmetry, (include triangles, quadrilaterals, polygons, circles, cones, cylinders, rectangular prisms, and pyramids).

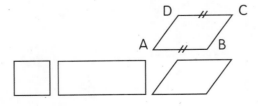

22. Answer: You would need 1,000 blocks.

Discussion

Tip: What do you know about computing volume?

Volume is length times width times height. The volume of a cube with 10 blocks on a side will be 10 x 10 x 10 = 1,000. You will need 1,000 blocks.

5.21 Determines a pair of numbers or the missing element of a pair when given a relation or rule, and determines the relation or rule given pairs of numbers.

23. Answer: 3,200 square feet

Discussion

Tip: How can you use the diagram to solve the problem?

You can count the number of whole squares inside the playground. There are 32 of them. Each represents a 10 ft by 10 ft square area, so each square has an area of 100 square feet. The area of the playground is 32 x 100, or 3,200 square feet.

Another strategy: The shape of the playground is a 60 x 60 feet square with a 20 x 20 ft corner removed. The 60 x 60 feet square has an area of 3,600 square feet. The 20 x 20 ft corner is 400 square feet. The area of the playground is 3,600 − 400, or 3,200 square feet.

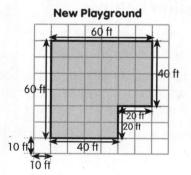

New Playground

60 ft
40 ft
60 ft
20 ft
20 ft
10 ft
40 ft
10 ft

5.11 Makes models of plane and solid figures and sorts and classifies these models according to characteristics such as number of sides, angles, vertices, faces, edges, tessellations, and lines of symmetry, (include triangles, quadrilaterals, polygons, circles, cones, cylinders, rectangular prisms, and pyramids).

5.15 Develops procedures to determine and compute perimeter, area, and volume of various geometric figures through concrete experiences with covering, filling, and counting.

24. Answer: 24 square units

Discussion

Tip: How can you draw the figure?

In each ordered pair, the first number shows how far to move to the right. The second number shows how far to move up. Graph all 4 points. Connect the dots. This rectangle is 6 units long and 4 units wide.

$A = l \times w$

$6 \times 4 = 24$ square units.

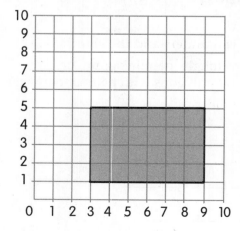

5.12 Uses ordered pairs of numbers to locate points on a grid or map and determine the ordered pair for a given point.

5.15 Develops procedures to determine and compute perimeter, area, and volume of various geometric figures through concrete experiences with covering, filling, and counting.

25. Answer: See discussion below.

Discussion

Tip: How will each 90-degree turn change the picture?

| Start | Rotate 1x | Rotate 2x | Rotate 3x | Rotate 4x |

A 90-degree turn will rotate the pattern one fourth of the way around. To check for accuracy, have students watch 1 shape (for example, the triangle) as it moves clockwise around the diagrams. After four rotations, you have the same figure you started with.

5.10 Determines line of symmetry and identifies geometric relations (e.g., parallel to, perpendicular to, intersect, horizontal, vertical, similar, congruent, flips, slides, and turns).

26. Answer: $300

Discussion

Tip: What do you need to find out first?

First you need to find out the total area of her living room.

12 feet x 15 feet = 180 square feet

Then you can divide to find the equivalent number of square yards.

180 ÷ 9 = 20

There are 20 square yards in 180 square feet.

Then you can multiply the number of square yards by the cost per yard to find the total cost.

20 x $15 per yard = $300

5.16 Determines and estimates amounts of money.

5.36 Adds, subtracts, multiplies and divides whole numbers up through four digits using both vertical and horizontal (rewrite vertically) presentation without and with regrouping (uses calculators for more laborious computations than these: four-digit addition and subtraction; multiplication and division of three- by two-digit numbers).

27. Answer: A total of eleven 8-foot lengths is needed.

Discussion

Tip: How many 4-foot sections are there?

Each two 4-foot sections can be combined to make one 8-foot length. Divide the number of 4-foot sections by two, then add the number of 8-foot sections to get the total number of 8-foot lengths.

The flower gardens each need 2 sets of 8-foot lengths, so the three gardens need six 8-foot lengths in all.

For the vegetable garden, first figure out the missing dimensions. They are all multiples of 4. The longest side is 12 feet long. It will need an 8-foot length and half of another one. Then note that there are two 8-foot sides plus three 4-foot sides. The vegetable garden will need five 8-foot lengths in all. Students might label the map of the garden to help see this.

To get the total number of 8-foot sections, add 6 + 5.
Some students may solve the problem by finding the sum of the perimeters of the gardens.

$$16 \text{ ft} + 16 \text{ ft} + 16 \text{ ft} + 40 \text{ ft} = 88 \text{ ft}$$

Each fence section is 8 feet, so they need 88 ÷ 8, or 11 sections.

Order Form	Number of 8-Foot Sections
Flower Garden	6
Vegetable Garden	5
Total	11

5.11 Makes models of plane and solid figures and sorts and classifies these models according to characteristics such as number of sides, angles, vertices, faces, edges, tessellations, and lines of symmetry, (include triangles, quadrilaterals, polygons, circles, cones, cylinders, rectangular prisms, and pyramids).

5.15 Develops procedures to determine and compute perimeter, area, and volume of various geometric figures through concrete experiences with covering, filling, and counting.

5.21 Determines a pair of numbers or the missing element of a pair when given a relation or rule, and determines the relation or rule given pairs of numbers.

5.33 Collects and organizes data into tallies, charts, and tables; determines appropriate scale and constructs bar graphs.

28. Answer: See discussion below.

Discussion

Tip: How are the sizes of the squares related?

To find perimeter you can add the lengths of all the sides or use the formula $2w + 2l = P$.

Kind of Square	Perimeter	Area
1 x 1 square	4	1
2 x 2 square	8	4
3 x 3 square	12	9
4 x 4 square	16	16
5 x 5 square	20	25
6 x 6 square	24	36

Possible relationships students will describe:

As the size of the square increases, the perimeter increases.

The perimeter of a square is 4 times one side of the square.

A square with n units per side will have a perimeter of $4n$.

A square with n units per side will have an area of n^2.

5.15 Develops procedures to determine and compute perimeter, area, and volume of various geometric figures through concrete experiences with covering, filling, and counting.

5.21 Determines a pair of numbers or the missing element of a pair when given a relation or rule, and determines the relation or rule given pairs of numbers.

5.33 Collects and organizes data into tallies, charts, and tables; determines appropriate scale and constructs bar graphs.

Georgia Quality Core Curriculum Objectives

5.3 Use estimation strategies (such as front-end, clustering, rounding, or reference point) to predict computational results of whole numbers, fractions, mixed numbers, and decimals.

5.7 Relates a fraction to a part of a whole, a part of a set, and a point on a number line; uses models to determine equivalent fractions. Uses fractions with denominators of 2, 3, 4, 5, 6, 10, 16, or 100.

5.8 Expresses an ordered pair of numbers as a ratio.

5.10 Determines line of symmetry and identifies geometric relations (e.g., parallel to, perpendicular to, intersect, horizontal, vertical, similar, congruent, flips, slides, and turns).

5.14 Uses customary and metric units to measure length, capacity/volume (use liquid and dry units), weight/mass, elapsed time and temperature (include measuring length to the nearest quarter inch, nearest millimeter and temperature below freezing).

5.15 Develops procedures to determine and compute perimeter, area, and volume of various geometric figures through concrete experiences with covering, filling, and counting.

5.24 Identifies needed information and selects the steps necessary to solve multi-step word problems.

5.25 Solves one-, two-, and three-step word problems related to all appropriate fifth grade objectives including those presented orally and in writing; those in charts, tables, and graphs; and those with extraneous or insufficient information.

5.30 Interprets and draws conclusions from charts, tables, and graphs (e.g., pictographs, and circle graphs).

1. Answer: D 32

Discussion

Tip: Consider how the lengths of the sides of the two figures are related.

Students should recognize that the figures are the same shape, which makes them similar. To help the students realize that the length of one side of the smaller figure is half the length of the same side in the larger figure, ask:

What is the length of the top of the smaller figure? the larger figure? How are these two numbers related? What is the length of the bottom of the smaller figure?

Item Numbers	Georgia QCC Objectives
1. D	5.10
2. H	5.10
3. B	5.8
4. F	5.8
5. D	5.8, 5.25
6. J	5.8
7. A	5.7
8. J	5.14, 5.25
9. C	5.15, 5.25
10. H	
11. D	5.24
12. H	
13. C	5.14
14. G	5.25
15. B	5.3
16. H	5.3
17. B	5.25
18. G	5.30

© Harcourt

19. Answer: No. The distance is 325 miles.

Discussion

Tip: How can you use the scale to help you solve the problem?

You can write the mixed number as a decimal number and then write a ratio to solve the problem:

$$3\frac{1}{4} = 3.25$$

According to the scale, the ratio of inches to miles on the map is 1 to 100, or $\frac{1}{100}$. So, 3.25 inches on the scale corresponds to 3.25 x 100 miles:

3.25 inches x 100 miles per inch = 325 miles

So, the actual trip distance is 325 miles.

5.8 Expresses an ordered pair of numbers as a ratio.

20. Answer: Yes. Both show a 3 to 1 ratio.

Discussion

Tip: How can you determine if ratios are equivalent?

To determine if a 6 to 2 ratio and a 12 to 4 ratio are equivalent, first write the ratios as fractions. Then write the fractions in simplest form.

$$\frac{6}{2} = \frac{3}{1}$$
$$\frac{12}{4} = \frac{3}{1}$$

Both fractions show a 3 to 1 ratio. Each team lost 1 game for every 3 games that they won.

5.7 Relates a fraction to a part of a whole, a part of a set, and a point on a number line; uses models to determine equivalent fractions. Uses fractions with denominators of 2, 3, 4, 5, 6, 10, 16, or 100.

© Harcourt

21. Answer: No

Discussion

Tip: How can you determine if two numbers have a 2 to 1 ratio?

To find the solution, write the number of flower stickers and the number of animal stickers as a fraction. Write the fraction in simplest form to determine if it is a 2 to 1 ratio.

$$\frac{108}{36} = \frac{3}{1}$$

Sarah has a 3 to 1 ratio of flower stickers to animal stickers.

5.8 Expresses an ordered pair of numbers as a ratio.

22. Answer: 25

Discussion

Tip: How can you use ratios to find a missing length in similar figures?

To find the missing length, first find the ratio of the lengths of the matching sides. The ratio of the length $\overline{AB}$ to the length $\overline{AC}$ is $\frac{20}{50}$.

$$\frac{20}{50} = \frac{2}{5}$$

Apply this 2 to 5 ratio to the small triangle. If the length of $\overline{DE}$ is 10, then the length of $\overline{DF}$ is 25. Since $\frac{10 \div 5}{25 \div 5} = \frac{2}{5}$, the ratios of the lengths of the matching sides are equivalent.

5.8 Expresses an ordered pair of numbers as a ratio.

23. Answer: $22.50

Discussion

Tip: How do you find a percent of a number?

Use decimals. Write 10% as a decimal: 0.10. To find 0.10 of $25, multiply:

$$\begin{array}{r} \$25 \\ \times\ 0.10 \\ \hline \$2.50 \end{array}$$

So, subtract $2.50 from $25 to find the reduced cost of the blouse, or $22.50.

24. Answer: 50%
Discussion

Tip: What should be the sum of all the percents? How will this information help you solve the problem?

The whole is represented by 100%. Add the given percents and subtract that sum from 100%. In Mr. Wolfe's class 50% of the students have at least 1 pet, because 25% + 15% + 10% = 50%. The sum of the percents should be 100%. This leaves 50% of his students who do not have any pets.

5.24 Identifies needed information and selects the steps necessary to solve multi-step word problems.

25. Answer: 375 swimmers
Discussion

Tip: How can you use a ratio table to help you solve the problem?

To find the number of swimmers in the water park, you can make a ratio table. Use the data given to make the table: there is 1 lifeguard for every 15 swimmers and there are 25 lifeguards on duty. So, there are 375 swimmers at the water park.

Number of Lifeguards	1	2	3	25
Number of Swimmers	15	30	45	375

5.8 Expresses an ordered pair of numbers as a ratio.

5.30 Interprets and draws conclusions from charts, tables, and graphs (e.g., pictographs, and circle graphs).

26. Answer: 13.5 mi
Discussion

Tip: How can you use a scale on a map to help you find the actual distance?

Ben measured the distance to be 4.5 in. Use the scale given to find the actual distance.

$$\frac{1 \times 4.5}{3 \times 4.5} = \frac{4.5}{13.5} \begin{array}{l} \text{map distance} \\ \text{actual distance} \end{array}$$

Ben travels 13.5 miles to go to volleyball practice.

5.8 Expresses an ordered pair of numbers as a ratio.

© Harcourt

27. Answer: See below.

Discussion

Tip: How can you compare the fractions and percents?

Using the survey results, you can determine the fraction of the students who answered each response. Place the number of students who answered the response over 100.

$\frac{5}{100}$ = watch 0 hours of television a week.

$\frac{35}{100}$ = watch 1–6 hours of television a week.

$\frac{40}{100}$ = watch 7–12 hours of television a week.

$\frac{20}{100}$ = watch 13 or more hours of television a week.

Write each of the fraction values above as a percent.

| 0 hours | 5% | 1–6 hours | 35% |
| 7-12 hours | 40% | 13+ hours | 20% |

So watching 7–12 hours of television a week is the most popular and watching 0 hours a week is the least popular.

Discussion

Tip: How do you use percents to help create a circle graph?

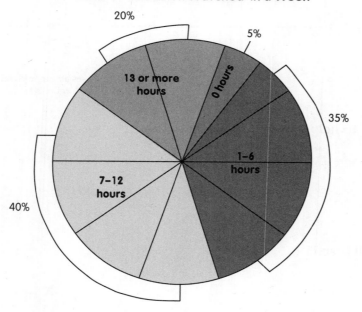

Hours of Television Watched in a Week

5.30 Interprets and draws conclusions from charts, tables, and graphs (e.g., pictographs, and circle graphs).

5.33 Collects and organizes data into tallies, charts, and tables; determines appropriate scale and constructs bar graphs.

© Harcourt

28. Answer: See below.

Discussion

Tip: How can you use a ratio to help you solve the problem?

The scale length and width can be found by using the scale 2 ft = 1 in. The actual length of his room is 12 ft. This means that the length of his room on the scale drawing is 6 in. The actual width of his room is 16 ft. This means that the width of his room on the scale drawing is 8 in.

Object	Measurement	Scale Measurement
Bed	6 ft X 8 ft	2 in. X 3 in.
Dresser	4 ft X 2 ft	2 in. X 1 in.
Desk	6 ft X 4 ft	3 in. X 2 in.

Discussion

Tip: How can you determine if the scale drawing will fit on the paper? What information do you need to use?

If Dan chooses to draw his scale drawing on a piece of paper that is $8\frac{1}{2}$ in. x 11 in., he will have enough room on the paper. The scale drawing of his room will only be 6 in. x 8 in.

5.4 Adds, subtracts, and multiplies fractions and mixed numbers with like and unlike denominators, using vertical and horizontal presentation.

5.30 Interprets and draws conclusions from charts, tables, and graphs (e.g., pictographs, and circle graphs).

5.33 Collects and organizes data into tallies, charts, and tables; determines appropriate scale and constructs bar graphs.

5.36 Adds, subtracts, multiplies and divides whole numbers up through four digits using both vertical and horizontal (rewrite vertically) presentation without and with regrouping (uses calculators for more laborious computations than these: four-digit addition and subtraction; multiplication and division of three- by two-digit numbers).